C0-DMS-510

Walter Klingmüller (Ed.)

Risk Assessment for Deliberate Releases

The Possible Impact of Genetically Engineered Microorganisms on the Environment

With 37 Figures

Springer-Verlag
Berlin Heidelberg New York
London Paris Tokyo

Professor Dr. WALTER KLINGMÜLLER
Lehrstuhl für Genetik, Universität Bayreuth
Universitätsstraße 30
8580 Bayreuth, FRG

TP
248.6
.R57
1988

WIDENER UNIVERSITY
WOLFGRAM
LIBRARY
CHESTER, PA.

DISCARDED
WIDENER UNIVERSITY

ISBN 3-540-18930-0 Springer-Verlag Berlin Heidelberg New York
ISBN 0-387-18930-0 Springer-Verlag New York Berlin Heidelberg

This work is subject to copyright. All rights are reserved, whether the whole or part of the material is concerned, specifically the rights of translation, reprinting, re-use of illustrations, recitation, broadcasting, reproduction on microfilms or in other ways, and storage in data banks. Duplication of this publication or parts thereof is only permitted under the provisions of the German Copyright Law of September 9, 1965, in its version of June 24, 1985, and a copyright fee must always be paid. Violations fall under the prosecution act of the German Copyright Law.

© Springer-Verlag Berlin Heidelberg 1988
Printed in Germany

The use of registered names, trademarks, etc. in this publication does not imply, even in the absence of a specific statement, that such names are exempt from the relevant protective laws and regulations and therefore free for general use.

Printing: Druckhaus Beltz, Hemsbach/Bergstr., Bookbinding: J. Schäffer GmbH & Co. KG., Grünstadt
2131/3130-543210

Preface

In 1986, the European Community launched its Biotechnology Action Programme (BAP) with the conclusion of 270 research contracts with laboratories in member states.

This programm covers a wide range of activities involving biological materials and processes in which new cellular and molecular methods, coupled with the application of information technologies, are expected to generate unprecedented technological progress. It concentrates on developing the most promising techniques in the fields of enzyme and protein engineering, genetic engineering, cell culture technology and culture collection, bio-informatics, "in vitro" pharmacological and toxicological screening, and risk assessment.

The risk assessment sector, considered to be of minor importance until a few years ago, is now emerging as a focal point of interest in biotechnology and will certainly be expanded considerably in the future programme. The research activities carried out in this area and supported by BAP, although small in number, are very significant and are related to both the containment and the release problems.

Genetically labelled Rhizobium is utilized as a model system to assess the extent of gene transfer to other members of the soil flora. With the aim of testing under different environmental and soil conditions, a common experiment has been started in experimental plots at Rothamsted Experimental Station, INRA Laboratory of Dijon and University of Bayreuth.
For the development of Baculoviruses to be used in the control of agricultural and forest pests, risk assessment analysis of genetically altered insect viruses is needed. A project involving field trial testing is carried out jointly by the NERC

Institute of Virology of Oxford and the BBA Institute of Darmstadt.
Novel methods and testing procedures to detect the escape of microorganisms from large scale bioprocessing and to validate the integrity of process components and unit operations were jointly studied by the Warren Spring Laboratory and the TNO Institute.

On October 26th to 28th the first meeting of all BAP-contractors working in the sector of risk assessment took place in Bayreuth, FRG. The meeting was supported by the European Community. It had been opened to the participation of many other European laboratories active in the same field, so that the various research approaches could be debated and the different views about the important problem of safety evaluation in biotechnology could be exchanged. The meeting was at the same time the first official international meeting on this topic in the FRG. It was hosted by the Genetics Department of Bayreuth University, with Professor Walter Klingmüller as local organizer.

A broad spectrum of topics was covered in the formal reports, including, among others, the Rhizobium and the Baculovirus experiments, the transfer of plasmids in the soil, the degradation of pollutants by bacteria, the problem of genetic stability and horizontal gene transfer after releases, the problem of possible biological containment, and that of containment in bioprocessing. Also the important problem of regulating the release of genetically manipulated organisms was covered at the meeting, with special attention to the experience in the USA, the United Kingdom and the Federal Republic of Germany. These reports are collected in full in this book.

In the free discussions of the meeting, a number of controversial issues were raised, viz. the definition of the terms "release" and "genetically engineered or altered", the need for guidelines or for a case by case evaluation, or for both, and the difference in attitude of industrial companies and university research towards releases and towards their funding and the question whether certain "safe" species or strains could be exempted from future regulations. Also a number of methodological problems were discussed lucidly, e.g.: how to standardize microcosm systems, and what use this would be, how to improve the detection level of released microorganisms, and how to achieve screening of bacteria and DNA in the soil economically.

These remarks will make clear, what a wide range of topics was dealt with. It is easy to conceive risks everywhere, in a general way, as for all other modern technologies. In contrast it will be difficult, or even impossible to demonstrate a complete absence of such risks. Sensible experimental work is needed urgently on a wide scope to dig out the facts and data that can put considerations and calculations on a solid basis. However, the field is in an extreme sense interdisciplinary and thus demanding, the gaps in knowledge and the research needs have to be precisely identified, suitable approaches have to be agreed upon. In addition to the hard work at the bench or even in the field, intellectual exchange, conversation, critical evaluation on the basis of mutual specific knowledge and experience are therefore crucial. The chance for such activities was offered in Bayreuth and was used by all participants. It is hoped that this book will help to promote the knowledge of facts as opposed to speculation.

W. Klingmüller

CONTENTS

List of Participants

Alef, K.	Bayreuth	Germany
Alexander, M.	Ithaca	USA
Anderson, P.	Copenhagen	Denmark
Backhaus, H.	Neuherberg	Germany
Beck, Th.	München	Germany
Beringer, J.E.	Bristol	England
Bertazzoni, U.	Pavia	Italy
Beyse, J.	Bayreuth	Germany
Brandmüller, E.	Bayreuth	Germany
Bricknell, D.	Brussels	Belgium
Buhk, H.J.	Berlin	Germany
Cowey, K.	London	England
Denarié, J.	Castenat-Tolosan	France
Döhler, K.	Bayreuth	Germany
Dums, F.	Bayreuth	Germany
Dwyer, D.	Genève	Switzerland
Economidis, I.	Brussels	Belgium
Elsas van, J.D.	Wageningen	Netherlands
Entwistle, P.F.	Oxford	England
Filip, Z.	Langen	Germany
Gadkari, D.	Bayreuth	Germany
O'Gara, F.	Cork	Ireland
Griffith, C.	Cardiff	England
Häusser, E.	München	Germany
Henschke, R.	Braunschweig	Germany
Herrmann, R.	Heidelberg	Germany
Heynen, C.	Wageningen	Netherlands
Hirsch, P.	Harpenden	England
Hooper, S.	Genève	Switzerland
Huber, J.	Darmstadt	Germany
Kale, N.	Pune	India
El-Khawas, H.	Bayreuth	Germany
Klemm, P.	Copenhagen	Denmark
Klingmüller, W.	Bayreuth	Germany

Kreutzer, R.	Bayreuth	Germany
Lange, P.	Bonn	Germany
Leaver, G.	Stevenage	England
Lotz, W.	Erlangen	Germany
Lorenz, M.	Oldenburg	Germany
Lugtenberg, B.	Leiden	Netherlands
Maat, C.	Wageningen	Netherlands
Mahler, J.L.	Copenhagen	Denmark
Milewski, E.	Washington	USA
Moawad, H.	Cairo	Egypt
Molin, S.	Copenhagen	Denmark
Nücken, E.	Braunschweig	Germany
Nuti, M.	Padova	Italy
Pelsy, P.	Paris	France
Rousseau, I.	Zeist	Netherlands
Ruckdäschel, E.	Bayreuth	Germany
Sarvas, M.	Helsinki	Finnland
Schilf, W.	Bayreuth	Germany
Schmidt, F.	Braunschweig	Germany
Schumann, W.	Bayreuth	Germany
Simon, R.	Bielefeld	Germany
Spokes, J.	Harpenden	England
Stewart, I.W.	Stevenage	England
Stolp, H.	Bayreuth	Germany
Stumpf, F.	Bayreuth	Germany
Undorf, K.	Darmstadt	Germany
Wackernagel, W.	Oldenburg	Germany
Wagner-Döbler, I.	Braunschweig	Germany
Werner, C.	Bayreuth	Germany

A MICROBIAL ECOLOGIST LOOKS ONCE AGAIN AT RISK ANALYSIS

Martin Alexander

Laboratory of Soil Microbiology, Department of Agronomy, Cornell University, Ithaca, NY 14853 USA

- Every innovative technology should be unquestionably accepted. Because the technology is novel, it is impossible to predict the consequences.
- This technology I am helping to develop is the greatest thing that has happened since man (or woman) has appeared on this planet. That new technology will likely have cataclysmic consequences and may kill many people, upset natural ecosystems and greatly modify the biosphere.
- Because the opponents of this technology do not understand the full details of the discipline, they should not be listened to. Because the proponents of the technology do not understand the implications for public health, food production or natural ecosystems, they are not the individuals whose opinions should be heard in assessing its consequences.
- No evidence of a problem has been unearthed to date, and therefore there is no problem with this technology. Evidence of the deleterious effects of nearly all technologies have required their widespread use or modification, and therefore we still may expect dire consequences.

Risk Assessment for Deliberate Releases
Edited by W. Klingmüller
© Springer-Verlag Berlin Heidelberg 1988

One of each of these pairs of statements is typical of the responses of many individuals to almost any new development in medicine, agriculture, industry or other human endeavors. In the debate about radically new activities of society, industry or agriculture, both viewpoints are commonly expressed. The disinterested observers of the introduction of atomic energy, antibiotics in human medicine, the establishment of large factories and the advent of the industrial revolution, the use of dynamite, the adoption in agriculture of pesticides, the burning of coal or municipal wastes and a host of other technologies could have been, or in fact were and still may be, bombarded by the same types of comments.

This is the dilemma of the risk analyst. That individual, often a scientist, although that is questioned by many, is not a proponent or opponent of one or another technology. The risk analyst is attempting to provide an objective evaluation, at least as objective as possible in the absence of a large data base, of whether there is in fact a risk and to determine the factors involved in that risk. The evaluation is not designed to propose or oppose a technology but rather to assess it. By such assessments, it is assumed that society will have a better way of determining which courses of action it should pursue and which it should avoid, and the information should be useful to those specialists who are concerned with the management of risks, that is, the development of procedures to reduce the risk and avoid the possible deleterious consequences. Risk analysis is not something that should be pursued, however, by active participants in the technology that is being evaluated because objectivity is then, legitimately, subject to question. Moreover, as the author can attest, it should not be pursued by individuals whose colleagues represent either the proponent or the opponent group or indeed, in my instance, come from both schools of thought. The final analysis may be characterized by objectivity, but the producer of the evaluation may be faced with a significant loss of amiable colleagues.

To the individual that participates in risk analyses, especially for regulatory agencies of government, or in reviewing technologies that have a reasonably long history, it is evident that the magnitude of risk is almost never evident initially. The proponents traditionally consider the benefits, often overemphasizing them, and either choose to ignore or refuse to admit or consider the possibility of deleterious consequences. To them, the probability of deleterious consequences is 0.0. Conversely, the early opponents of the technology, or those individuals who become aware of it as it becomes widespread, deem the new field of endeavor to be extremely hazardous, and they often fail to consider the many possible benefits to particular segments of society or to society at large. To many of these individuals, the probability of risk is 1.0. The industrial revolution with the subsequent harm done to air and water quality, the widespread development of the chemical manufacturing industry with the consequent environmental pollution, the introduction of the chlorinated hydrocarbon insecticides with the ultimate stress placed upon bird and fish populations, the introduction of antibiotics into chemotherapy and the harm done to sensitive individuals, the use of atomic energy to provide power and the widespread use of synthetic fertilizers and the resulting eutrophication represent but a few of the many examples that could be cited of technologies that have modest, major or enormous benefits, yet each has resulted, or has been perceived to result, in harm to lesser or greater numbers of people or to natural plant and animal populations. Each has a group of advocates who represent the technology and state emphatically its benefits. Each has its detractors that point, often with due cause, to the harm. Retrospective evaluations are frequently cited to justify one viewpoint or another, but the risk analyst concerned with the initial phases of a technology does not have the advantage of retrospective evaluations and must use prospective assessments. He must separate rhetoric and self-serving opinions from what is needed for proper assessments. Unfortunately, facts are rarely available for the prospective evaluation, and a variety of estimates, some good and others tenuous, must be employed to develop a useful approach.

With regard to genetically manipulated organisms, a variety of arguments have been advanced to buttress the divergent viewpoints. Some of these arguments are impressive, cogent and worthwhile. Others are interesting but, at best, fanciful. Still others represent attempts by the proponents or opponents to maximize the chance of widespread acceptance of their views. Again, the citing of good, tenuous or highly dubious arguments is characteristic of proponents or opponents of individual technologies.

A key issue is not whether a particular product of a technology, genetic engineering or other, is going to cause harm but what is the probability of any product creating undesirable effects. Careful consideration of the issues on a case-by-case basis and by unbiased reviewers will often allow a meaningful decision for a particular organism. However, a technology that has a realm of possibilities, a large number of modifications in potential applications, a variety of procedures to alter the technology or the organisms for particular purposes cannot be assessed so readily with a view to deciding on its safety. The issue then is not simply harm or no harm, as it would be for a particular genetically manipulated organism; for the technology as a whole, the issue is one of probability. Considering that many genes are ultimately transferable to many species, and that such genetic manipulation can be used in a variety of ways, it is naive to say that there is or is not a problem; that is, that the risk is 100 or 0 percent. In their infancies, other technologies did not show evidence of a lack of hazard. For example, the chemical industry developed for many years and had a variety of applications and uses before the dangers became evident. The problems were always characteristic of the industry, but they were not readily evident until new directions or wide uses of the chemical industry came to the fore. Furthermore, even with the best of safeguards, an adequate base of information and careful regulatory review, problems arise because of sloppiness on the part of engineers or commercial exploiters, inadequate care to control or the activites of unethical entrepeneurs. Probably every unregulated industry has had, and will continue to have, a variety of these problems, and a careful regulatory system provides controls that would not exist if there were no system of regulation.

Considering that a risk analysis requires information on two general matters, exposure and potential harm, an approach can be developed to separate out the various factors that need to be evaluated in determining risk. The probability of exposure requires that a product of a technology that may cause injury is released, that it persists, and that it be transported to a site containing the susceptible individuals. Such an analysis is characteristic of, for example, evaluating the risk from chemicals and air pollutants: will the harmful agent be released, will it persist or be modified or destroyed in time, is it subject to transport to sites at which susceptible individuals live, and finally will it have a deleterious influence? Considering that living organisms not only persist but multiply, the list of factors involved in such a risk analysis was modified to include the possibility of proliferation. Moreover, inasmuch as living organisms not only survive and multiply but are able to transmit genetic information to other species, gene transfer was also considered as one of the factors involved in the risk analysis. Thus, a presumably objective approach to risk analysis from genetically manipulated organisms should consider the probabilities of (a) release, (b) the survival of the organism following its release, (c) its multiplication, (d) the transfer of genetic information to other species, (e) the dissemination of the organisms and (d) a harmful effect that might ensue (1). Inasmuch as the specific concern is now with deliberately released organisms, the first factor, namely, the possibility of release, can be ignored: its probability is 1.0.

The author, as a microbial ecologist, is comfortable with evaluations of survival, multiplication and dispersal. This takes me away from risk analysis, an area in which I serve as a government advisor but not as a researcher, back to the types of investigations that we conduct. Much of our research has dealt with survival of bacteria, fungi, algae and protozoa in soils and waters. We have also conducted a small number of studies on the capacity of microorganisms to grow in natural ecosystems, especially around plant roots and in surface waters. Even without referring to the voluminous literature in these areas and our own modest contributions, it is clear that microorganisms introduced into ecosystems in which they are not native frequently die in a short period of time and no survivors remain after days or weeks; on the other hand, some of these alien species often endure and persist for many weeks, months and even years. Furthermore, the published literature, although not large, indicates that growth of species in ecosystems in which they are not indigenous sometimes occurs, and the introduced organism may occasionally reach a considerable cell density or biomass. Although the alien rarely grows and survives to the extent that it becomes a significant component of the invaded ecosystem, survival and multiplication can be of sufficient duration to pose a risk to other species that might be susceptible to the harm done by the introduced organism.

A large amount of information also exists on microbial dispersal. That knowledge comes largely from studies of epidemiology of human and animal diseases and also from assessments of the transport of plant pathogens. Those studies or field monitoring activities indicate that some organisms move tremendous distances, other are dispersed to modest degrees whereas still others fail to move from the site of their initial introduction.

However, the available information on survival, multiplication and dispersal is not sufficiently detailed or abundant to allow for predictions of the capacity of previously untested microorganisms to survive, multiply or disperse. The research has largely been descriptive and was not designed to establish generalizations to allow for predictions of behavior of as yet untested species or other taxa.

Reasonable procedures now exist to assess the survival and fate of genetically manipulated microorganisms that are likely to be released in the future. The continued research in this area is important and worthwhile, and it will allow a better definition of the issues that appear in risk analysis and the final evaluation of the safety or possible dangers of individual products of genetic manipulation. On the other hand, it is still totally unclear how to evaluate deleterious effects, particularly with regard to species that are not of economic or public health concern. Specialists are abundant in human and veterinary medicine and in plant pathology, but these specialists deal with only a very small number of species. It is likely, should a genetically manipulated organism of some sort do harm, that the injury will be to some species that does not happen to be on the small list of test species for which methods have been developed by plant pathologists and veterinary researchers. Indeed, the likelihood of harm to a particular species, from a purely statistical viewpoint, is one divided by the number of species. From the viewpoint of the biosphere as a whole or of particular ecosystems, there are no testing procedures currently available for assessing harm to most species, and thus no way of evaluating, under controlled circumstances, whether a particular genetically manipulated organism will or will not be harmful to one of those species. Furthermore, the cost of evaluating potential injury to higher plants or, even more so, to humans and economically important animals is so great that the evaluations of risk and the review of particularly genetically engineered organisms will become so expensive as to markedly inhibit the development of an important area of scientific and technological endeavor. Hence, although continued research on survival, multiplication, gene flow and dispersal is important, the greatest deficiency in actual assessments of hazard is the inability to predict, test, and evaluate the deleterious effects of genetically manipulated organisms on natural populations and natural communities.

From the viewpoint of the risk analyst, it is foolhardy to assume certainty---certainty of harm or the absence of harm. The risk remains unknown. However, it is possible to delineate the various areas in which knowledge is required in order to provide meaningful risk analyses. That same delineation of needs also provides a basis to evaluate individually genetically manipulated organisms prior to their release. The availability of this information will reduce the uncertainties on risk quite considerably. That knowledge will allow for the development of reliable means for assessing the safety of individually genetially manipulated organisms. In this way, society will be able to gain benefits from the revolution in molecular genetics and have an increasingly small probability of deleterious effects.

References

1. Alexander, M. (1985). Issues in Science and Technology, 1 (3), 57-68.

RHIZOBIUM LEGUMINOSARUM AS A MODEL FOR INVESTIGATING GENE TRANSFER IN SOIL

Penny R. Hirsch and John R. Spokes
AFRC Institute of Arable Crops Research, Rothamsted Experimental Station,
Harpenden, Herts AL5 2JQ, UK

Joint project on: Assessing the risks involved in the release of genetically manipulated microorganisms (partners: W. Klingmüller and K. Döhler, Bayreuth; N. Amarger, Dijon).

Summary: A conjugative symbiotic plasmid of *Rhizobium leguminosarum* bv. *viceae* was marked with transposon Tn*5* and the strain was used to inoculate peas in the field. The Tn*5* marker could be readily identified in rhizobia re-isolated from root nodules. The experiment was designed as a model system to investigate gene transfer in the environment between introduced and native strains of bacteria. The rationale behind the experiment and details of the strain construction and environmental release as well as preliminary results, are reported.

Keywords: environmental release, risk assessment, gene transfer, *Rhizobium*, transposon

Introduction

The techniques of genetic manipulation have already been applied to agriculturally-important bacteria, and will certainly lead to the production of improved strains. These microbes should be rigorously tested in laboratory conditions before deliberate release into the environment can be considered. However, in the field they may exchange genes with native strains producing novel hybrids with properties that are not easy to predict. The lack of information on gene transfer between bacteria in the environment makes it difficult to assess the risk of the formation of undesirable hybrid strains.

The ubiquitous soil bacteria belonging to the genus *Rhizobium* form nitrogen-fixing symbiosis in root nodules of certain leguminous plants, the species or biovar of *Rhizobium* being named after the plant species with

Risk Assessment for Deliberate Releases
Edited by W. Klingmüller
© Springer-Verlag Berlin Heidelberg 1988

which it associates. Rhizobia have been used as agricultural inoculants for legume crops for over one hundred years and have never been reported to be harmful to man, plants, or the environment. The first pure cultures of Rhizobium isolated from root nodules and produced commercially were called "Nitragin", patented by Nobbe and Hiltner in 1895, and manufactured by Meister, Lucius and Brüning at Höchst am Main in Germany in the late nineteenth century. Today throughout the world, 20 million hectares of legumes are inoculated with Rhizobium each year and many scientists are working towards improving the symbiotic performance of rhizobia. It is probable that proposals will be made in Europe in the near future to test rhizobia improved by genetic manipulation in the field, and so any information on gene transfer between rhizobia in the environment will be highly relevant. Rhizobium has several other advantages for such studies: the genetics of various species have been studied in detail in the laboratory and conjugative plasmids have been identified; large numbers are present in most field soils, and unlike most soil bacteria they can be readily re-isolated from soil via the root nodules of plants with which they form symbiotic associations.

Rhizobium strains contain two or more large plasmids and in most cases genes for nitrogen fixation and specific host plant nodulation are carried on one, the symbiotic plasmid or pSym. Transfer of plasmids between different biovars and species has been observed both on agar and during plant infection in laboratory conditions. However, to monitor this process efficiently it is necessary to introduce a selectable marker such as an antibiotic resistance gene. The transposon Tn5 which encodes resistance to kanamycin (Km) and neomycin (Nm) has frequently been used to mark Rhizobium plasmids. Its origins are uncertain since it can express transposition and antibiotic resistance properties in a wide range of Gram-negative bacteria, but it was first identified from a clinical isolate of Klebsiella pneumoniae from Washington D.C. (1). The properties of Tn5 have recently been extensively reviewed (2). It is capable of transposition to numerous sites in chromosomal and plasmid DNA and has been a valuable tool for the genetic manipulation of many organisms. In addition to the Km^R and Nm^R determinants it confers resistance to bleomycin and in some bacteria it also confers streptomycin (Sm) resistance (3, 4). The entire genome of Tn5 has been sequenced (5) and it has been shown to have two 1.5 kb terminal inverted repeats encoding transposition functions and a central unique 2.7 kb region containing the antibiotic resistance genes. The antibiotic resistance genes and the unique DNA sequence make Tn5 a very useful selectable marker, and the ability to transpose means that it can be introduced into cells on DNA replicons that cannot themselves be maintained. Tn5 is thus a suitable marker for detecting transient genetic interactions, and since during the past 15 years no adverse effects of exposure have been reported by any of the numerous groups who work with it, it may also be considered to be a safe marker.

To investigate gene transfer in the environment, we have marked a conjugative symbiotic plasmid in an R. leguminosarum bv. viceae strain with Tn5, and used it as an inoculant for peas in the field. For a control where no positive selection for R. leguminosarum bv. viceae would be exerted, wheat and barley were also inoculated. In the first year, nodules from the peas have been screened for the presence of the inoculant and other R. leguminosarum bv. viceae strains carrying Tn5. In subsequent years the plots will be planted with a range of legumes (peas, beans and clover) to allow sampling for the presence of Tn5 in populations of R. leguminosarum bv. viceae, bv. phaseoli and bv. trifoli.

Because Tn5 came originally from a different bacterial family, the Rhizobium inoculant strain is classed as an interspecific hybrid. The UK Government Advisory Committee on Genetic Manipulation (ACGM) was consulted, but stated that they had no objections to the field experiment proceeding.

Materials and methods

Conditions for culturing E. coli and Rhizobium, performing crosses and screening have been described elsewhere (6). Rhizobia were isolated from root nodules on yeast-mannitol agar containing Congo Red dye (7) supplemented with 100 $\mu g\ ml^{-1}$ cyclohexamide to inhibit fungi. Nodules were crushed after washing in 95% ethanol, sterilizing for 1 minute in "Chloros" (11% available Cl) then rinsing with five changes of sterile distilled H_2O. Antibiotics from Sigma were used at the following concentrations in $\mu g\ ml^{-1}$ (abbreviation of antibiotic also given): Streptomycin (Sm) - 200; Rifampicin (Rif) - 50; Neomycin (Nm) - 50; Kanamycin (Km) - 50; Gentamicin

(Gm) - 20, Chloramphenicol (Cm) - 50, Spectinomycin (Sp) - 100. Colony and gel blots, ^{32}P-labelled DNA for probes and blot/probe hybridization conditions were essentially as described in Maniatis et al. (8). Peat-based inoculant for field application was prepared by NPPL (AGC, Cambridge, UK). Plants used in the field were: Peas- cv. Progreta (a genetically homogenous variety resistant to Pea Blight races 1, 2, 3, 5); Spring Barley - Klaxon; Spring wheat - Wembley; Chick peas unknown cultivar of a commercial variety.

Table 1. Bacterial strains and plasmids

Bacteria	Relevant characteristics	Source/ reference
E. coli		
JI1830	contains pJB4JI	(9, 10)
R. leguminosarum bv. viceae		
JI248	field isolate	(11)
RSM2000	Sm^R derivative of JI248	This lab.
RSM2001	Rif^R derivative of RSM2000	"
RSM2002	Rif^R derivative of RSM2000	"
RSM2004	RSM2001 with Tn5 insertion on pRL1JI	"
R. leguminosarum bv. phaseoli		
JI8002	field isolate	(12)
JI8400	JI8002 cured of psym	(12)
8400 Sp^RCm^R	derivative of JI8400 made sequentially Sp^R and Cm^R	This lab.
Plasmids		
pKan2	contains central Hind III fragment from Tn5 - used as probe for Tn5	(13)
RP4::Mu::Tn7	contains Mu cts	(9)
pJB4JI	suicide vector for Tn5	(9, 10)
pRL1JI	symbiotic plasmid from JI248	(11)

Construction of Inoculant Strain RSM2004

R. leguminosarum bv. viceae strain JI248 was isolated from V. faba root nodules in Norfolk, UK. It was found to contain a conjugative plasmid, pRL1JI, that carried symbiotic and bacteriocinogenic genes and could transfer to other Rhizobium strains at a frequency of about 10^{-2} per parent (11, 14). There are at least six large plasmids in JI248 ranging in size from about 80-400 Md and pRL1JI, at 130 Md, is the second smallest (15). Two spontaneous Rif^R mutants were isolated from a Sm^R derivative of JI248 (Table 1), and since the mutations map to the chromosome they are convenient markers for strain recognition. These derivatives were crossed with E. coli strains JI1830 that carried a "Suicide vector" plasmid containing Tn5, pJB4JI (9, 10). This plasmid is a co-integrate of bacteriophage Mu with an in vivo recombinant - IncP1 plasmid pPH1JI, and Tn5 is inserted in the Mu moeity. It can replicate stably in E. coli and transfer to Rhizobium but cannot replicate in this host: if Tn5 is selected by plating Rhizobium transconjugants on medium supplemented with Nm or Km, the majority of resistant colonies are found to have Tn5 stably inserted in their DNA and to have lost the vector, pJB4JI. Twenty such derivatives from crosses between JI1830 and RSM2001, and RSM2002 were isolated and shown to have lost the Gm^R marker carried on pJB4JI. Plasmid gels confirmed that no extra bands were present in these strains, and gel blots hybridized to Tn5 probes showed that 3/20 RSM2001::Tn5 clones and 2/20 RSM2002::Tn5 clones contained Tn5 insertions on pRL1JI. Hybridization to Mu DNA (isolated from bacteriophage particles obtained by heat-induction of E. coli carrying RP4::Mu::Tn7) showed that no Mu sequences were present in these five strains. Insertion onto pRL1JI was verified by selecting for transfer of the Tn5 marker to non-nodulating strain $8400Sp^RCm^R$: Km^R transconjugants acquired the ability to form nitrogen-fixing root nodules on peas and had an extra plasmid band corresponding in size to pRL1JI. This indicated that no major symbiotic functions or transfer genes had been disrupted by Tn5 insertion and the bacteriocinogenic properties (14) were also found to be intact. Both the Rif^R strains RSM2001 and RSM2002 and their derivatives carrying Tn5 were found to be less competitive than the parent strain 248; forming only about 10% of the nodules when equal numbers were used to inoculate peas. However, in competition with two other field isolates one derivative, RSM2004, appeared to be able to form about 50% of root nodules and was therefore chosen for use in the field experiment.

Release of RSM2004

The field of the Rothamsted farm chosen for the experiment is a free-draining silty clay loam which has no public access, no surface water runoff and requires no drainage ditches. The soil contained between 10^3 and 10^4 R. leguminosarum bv. viceae per gm (methods according to ref. 7) and when 2.10^8 RSM2004 were applied as liquid inoculant to a pot containing 600 g soil and one geminating pea seed, it formed about 30% of the nodules. It was decided to use both a seed-coating inoculant which provided 10^6 rhizobia per seed to favour nodulation of the main root as it emerged, as well as a granular inoculant, spread along the drill furrows before planting the seeds to nodulate the later-appearing lateral roots. The granular inoculant provided 10^8 rhizobia per pea seed (10^{10} per m^2, approximately equal to 3 x 10^4 rhizobia per g soil). The indigenous rhizobia were less than 10^4 per g soil so the inoculant was in excess. [These calculations assume that the top 25 cm of soil contain the rhizobia and that under each m^2 of the field in question the top 25 cm of dry soil would weigh 280 kg (16)]. The amount of inoculant was over 5x the normal agricultural level. Although it is unusual to inoculate legumes when effective rhizobia are already present in significant numbers, it was necessary to give the inoculant strain numerical advantage since it was not especially competitive. Two plots 5 x 21 m were inoculated with the granular inoculant. One was sown with inoculant-coated peas and the other with a mixture of spring barley and wheat (the cereal seeds were not coated with inoculant) and the two strips were separated by a control plot of the same size planted with Chick peas inoculated with Rhizobium "cicer". The control was included to allow an estimation of movement of Rhizobium over the site, since Rhizobium "cicer" do not occur naturally in Rothamsted soil and the host plant can act as a sensitive monitor to detect these rhizobia in soil samples.

After inoculation the seed was sown in mid-May 1987 and the plots were covered with a net to keep out birds and surrounded by an electric fence to deter rabbits. Nodules from the peas were sampled at intervals to check for the presence of RSM2004 and for any indication of transfer of the Tn5 markers to other rhizobia. The site will be planted with peas, beans and white clover for the next two years to allow screening of the populations of *R. leguminosarum* bv. *viceae* bv. *phaseoli* and bv. *trifolii* respectively for the presence of Tn*5*. If this is still detected three years after inoculation, the site will be monitored indefinitely.

Results

Bacteria were sampled from pea root nodules and screened for Nm^R, Sm^R and Rif^R and for DNA homology to the Tn*5* DNA probe. Six weeks after planting, 20% of nodules from main roots and 3% from lateral roots contained rhizobia with all these markers, assumed to be RSM2004. At ten weeks only 2% of lateral root nodules appeared to contain RSM2004 (the main-root nodules were no longer intact due to pest damage or senescence). None of the rhizobia from 467 nodules, appeared to contain Tn*5* without the other markers (Sm^R Rif^R) of RSM2004. The final sampling of about 700 root nodules was performed 17 weeks after planting, and preliminary results indicated that about 10% of nodules contained Nm^R rhizobia with homology to Tn*5*: these are currently being screened for the presence of the other markers.

Discussion

The nodule occupancy in the field by RSM2004 in competition with native rhizobia was less than in comparable laboratory reconstructions, reinforcing the view that genetically altered rhizobia are often less well adapted to the field than the laboratory and may not survive well. However, a large number of the rhizobia containing Tn*5*, in total about 10^{12}, were released into the field, the number equalling that of the native *R. leguminosarum* strains and if gene transfer occurs at a significant frequency it should be apparent from the results during the next two years.

Acknowledgements

We thank Amanda Latham, Mary White and Daphne Gibson for their assistance, and John Jebb, John Day and Paul Williams for help and advice on production of the inoculant. This work was supported by a grant from the EEC Biotechnology Action Programme in Risk Assessment, and was carried out in collaboration with Noelle Amarger in Dijon, France, and Walter Klingmüller and Karl Döhler in Bayreuth, FRG.

References

1. Martin, C.M., Ikari, N.S., Zimmerman, J. and Waitz, J.A. (1971). J. Inf. Diseases 124 supplement S24-S29.
2. Berg, D.T.and Berg, C.M. (1983). Biotechnology 1, 417-435.
3. Genilloud, O., Garrido, M.C. and Moreno, F. (1984). Gene 32, 225-233.
4. Putnoky, P., Kiss, G.B., Ott, I. and Kondorosi, A. (1983). Mol. Gen. Genet. 191, 288-294.
5. Mazodier, P., Cossart, P., Girand, E. and Gasser, F. (1985). Nucl. Acids Res. 13, 195-205.
6. Wang, C.L., Beringer, J.E. and Hirsch, P.R. (1986). J. Gen. Microbiol. 132, 2063-2070.
7. Vincent, J.M. (1970). A Manual for the Practical Study of Root-Nodule Bacteria. Blackwell Scientific Publications, Oxford, UK.
8. Maniatis, T., Fritsch, C.F. and Sambrook, J. (1982). Molecular Cloning. A Laboratory Manual. Cold Spring Harbour Laboratory, USA.
9. Beringer, J.E., Beynon, J.L., Buchanan-Wolloston, A.V. and Johnston, A.W.B. (1978). Nature 276, 633-634.
10. Hirsch, P.R. and Beringer, J.E. (1984). Plasmid 12, 139-141.
11. Johnston, A.W.B., Beynon, J.L., Buchanan-Wolloston, A.V., Setchell, S.M., Hirsch, P.R. and Beringer, J.E. (1978). Nature 276, 634-636.
12. Lamb, J.W., Hombrecher, G. and Johnston, A.W.B. (1982). Mol. Gen. Genet. 186, 449-452.
13. Scott, K.F., Hughes, J.E., Gresshoff, P.M., Beringer, J.E., Rolfe, B.G. and Shine, J. (1982). J. Mol. Appl. Genet. 1, 315-326.
14. Hirsch, P.R. (1979). J. Gen. Microbiol. 113, 219-228.
15. Hirsch, P.R., Johnston, A.W.B., Brewin, N.J., Van Montagu, M. and Schell, J. (1980). J. Gen. Microbiol. 120, 403-412.
16. Hall, A.D. (1905). The Book of Rothamsted Experiments. John Murray, London.

GENETIC INTERACTION OF RHIZOBIUM LEGUMINOSARUM BIOVAR VICEAE WITH GRAM-NEGATIVE BACTERIA

Karl Döhler and Walter Klingmüller
Lehrstuhl für Genetik, Universität Bayreuth
Postfach 101251, D-8580 Bayreuth, FRG

Joint project on: Assessing the risks involved in the release of genetically manipulated microorganisms (partners: P.R. Hirsch and J.R. Spokes, Rothamsted, UK; N. Amarger, Dijon, France).

Summary: The genetically marked strain *Rhizobium leguminosarum* biovar *viceae* RSM2004, which contains the transposon 5 integrated in the conjugative symbiotic plasmid pRL1JI, was used for assessing the risks of releasing experiments. In the laboratory, transfer of pRL1JI to gram-negative bacteria and transposition of Tn5 were documented. After releasing of strain RSM2004 into the environment its survival and its ability to nodulate peas were measured. Plasmid and Tn5 transfer is also being checked and preliminary results are reported.

Keywords: risk assessment, environmental release, *Rhizobium*, soil bacteria, gene transfer, transposon 5

Introduction

One of the risks in releasing genetically manipulated organisms into the environment is that they might produce novel hybrids with the indigenous species, whose properties are extremely difficult to predict. Such approaches can therefore create problems of potential hazard to human, animal or plant health and of environmental impact in general. While there is some experience of risk assessment based on the evolution of novel traits in existing populations and the introduction of organisms to ecosystems, there is no experience as yet involving the introduction of genetically manipulated organisms. The risk assessment of such projects may involve many factors, reflecting the complexity of environmental interactions.

Risk Assessment for Deliberate Releases
Edited by W. Klingmüller
© Springer-Verlag Berlin Heidelberg 1988

This EC-joint project was designed to assess the risk of gene transfer from a genetically marked soil bacterium to other members of the soil microflora. As model organism a Rhizobium strain was chosen, because it is likely that in the future, symbiotically improved rhizobia will be produced using genetic manipulation to release them as inoculant for legume crops. Bacteria of the agronomically important genus Rhizobium are easy to select and identify by isolating them from root nodules of the leguminous plant with which they form symbiotic nitrogen-fixing associations. Rhizobia carry conjugative plasmids and transfer of such plasmids to other rhizobia has been demonstrated (1, 2, 3).

To investigate gene transfer in the environment under this EC-joint project a genetically marked strain of Rhizobium leguminosarum bv. viceae was used to inoculate - in conjunction with work in Rothamsted, UK and Dijon, France - an experimental plot at Bayreuth, FRG. We report data from laboratory studies connected to these investigations and first results from the actual releasing experiments.

Materials and methods

Bacterial strains and plasmids: Most of the bacterial strains used in the present study are listed in Table 1 (see 3 pages on), those chosen as recipients in mating experiments were naturally occuring nalidix acid (Nd) or carbenicillin (Cb) resistant derivatives. The genetically marked strain Rhizobium leguminosarum bv. viceae RSM2004 with the conjugative plasmid pRL1JI::Tn5 was constructed by Dr. P. Hirsch's group (3). Tn5 as probe for DNA-DNA hybridization was isolated from plasmid pSUP2021 (4).

Media and growth conditions: Complete medium for Escherichia coli, Enterobacter, Klebsiella and Pseudomonas was LB medium (5) and for Rhizobium tryptone yeast medium (TY) (6); minimal medium for Enterobacter (M56) was as described earlier (7) and minimal medium for Rhizobium (YM) was prepared according to Vincent (6). The following concentrations of antibiotics were used: Streptomycin (Sm), 200 μg/ml; Rifampicin (Rf) 50 μg/ml; Neomycin (Nm) 50 μg/ml; Kanamycin (Km) 50 μg/ml; Nalidixic acid (Nd) 50 μg/ml; Carbenicillin (Cb) 150 μg/ml.

Isolation of rhizobia: Rhizobia were isolated from root nodules on YM agar according to Vincent (6). Nodules were crushed after washing for 90 s in 95 % ethanol, sterilized 2-3 min with 6 % H_2O_2 and then rinsed ten times with sterile water. Direct isolation of strain RSM2004 from soil samples was performed by plating soil extracts on selective YM plates with Rf, Nm, and cycloheximide (75 µg/ml) to inhibit growth of fungi.

Plate-conjugation: Exponential phase cultures of strain RSM2004 as donor and various strains as recipients were mixed, the volume filled up to 2 ml and poured out on TY agar plates. After an overnight incubation at 30°C cells were washed from the plates and crossed out on selective agar plates (LB or M56) containing Km, Nm and the appropriate antibiotics for the recipient (Nd or Cb).

DNA biochemistry: Crude lysates of plasmid DNAs were prepared essentially according to Kado and Liu (8). Electrophoresis of crude lysates was carried out as described by Maniatis et al. (5). Plasmid pSUP2021 (4) was isolated by alkaline cell lysis (9). A Tn_5_ probe was obtained from the plasmid pSUP2021 by digestion with _HpaI_ endonuclease and preparative agarose gel electrophoresis. This fragment was labeled with biotinylated 11-dUTP by nick translation according to the instruction manual of Gibco-BRL (10). The labeled Tn_5_ was used for Southern blot analysis and colony hybridization (5). Conditions for DNA hybridization and detection were as described by Leary et al. (11) or, where modified, followed the specifications of the BRL manual (10).

Growth chamber experiments: Selective isolation of _R. leguminosarum_ bv. _viceae_ was done by inoculating soil extract to sterile pea plants in Erlenmeyer flasks according to Vincent (6). Pea plants were then grown in the growth chamber under light (16 h; approximately 100 µE) at 22°C and at 19°C in the dark (8 h). For mating experiments soil of the experimental field (loamy sand) was filled into pots. We inoculated a total number of about 3×10^9 cells of strain RSM2004 as donor and _Enterobacter agglomerans_ 339 as recipient, respectively. This corresponds to about 10^7 cells per g of soil (dry weight). Samples were taken by vigorous resuspending of about 1 g of soil from the pots in 5 ml of saline (0.9 % NaCl), diluting and plating on selective agar plates containing appropriate antibiotics and cycloheximide (75 µg/ml) to inhibit growth of fungi.

Field inoculation: The preparation of peat-based field inoculant, the conditions for releasing the strain RSM2004 to the experimental plot and the planting diagram was as described by Hirsch and Spokes (3).

Results

Transfer of pRL1JI in laboratory experiments

The conjugative symbiotic plasmid pRL1JI marked with Tn5 was reported to transfer from R. leguminosarum bv. viceae RSM2004 to other Rhizobiaceae in the range of 10^{-1} to 10^{-7} per parent. The plasmid can replicate in rhizobia and is stable in the examined exconjugants (1, 2, P.R. Hirsch, personal communication). In our laboratory we found transfer of kanamycin resistance from strain RSM2004 to some Enterobacter, in addition to E. coli (Table 1). Kanamycin and neomycin resistant exconjugants from the conjugation between strain RSM2004 and several strains of Enterobacter and E. coli were isolated and up to 40 colonies of each recipient strain were hybridized with biotinlabeled Tn5 probes. In contrast to the wild type strains more than 90 % of the exconjugants showed strong hybridization signals indicating that the majority of the obtained exconjugants received Tn5, coding for kanamycin resistance, from the donor via conjugative transfer of pRL1JI and were not naturally resistant mutants. The transfer rates of kanamycin resistance were about 10^{-6} to 10^{-7} for the E. coli strains and 10^{-6} to 10^{-8} for the Enterobacter agglomerans strains. However, at the present stage of our investigations we could not achieve transfer of kanamycin resistance from RSM2004 to some other strains of Enterobacter, to Pseudomonas or Klebsiella (Table 1). According to Johnston et al. (1) and Hirsch (2), plasmid pRL1JI, after transfer from RSM2004 to other rhizobia, is being replicated in the exconjugants. Therefore crude lysates of plasmid DNAs from Enterobacter and E. coli exconjugants were separated on agarose gel to check for the presence of plasmid pRL1JI, however with negative results. We then tested for the presence of Tn5 in the total DNA of the exconjugants. DNA from agarose gels was transferred to Nitrocellulose filter via

Table 1. Occurence of kanamycin resistant exconjugants in matings of _R. leguminosarum_ bv. _viceae_ RSM2004 with several other gram-negative bacterial recipients

Recipient	Exconjugants
Enterobacter agglomerans (12, 13)*	
strains 50, 233, 243, 307, 335, 339	yes
strains 234, 333, 334	n.o.
Enterobacter cloacae MF10 (12, 13)	n.o.
Escherichia coli	
strains HB101 (14), C600 (15), C2110 (16)	yes
Klebsiella pneumoniae UNF5023 (17)	n.o.
Pseudomonas	
aeruginosa PA05 (18), _stutzeri_ PSZ1 (19), _stutzeri_ ssp. (20)	n.o.
fluorescens 50090, _fluorescens_ 50091, _putida_ 50201 (21)	n.o.

n.o. : No exconjugants obtained
* : References of strains are given in brackets

the Southern blot method. The Southern filter was hybridized with biotin-labeled Tn_5_ (Figure 1). It was found, that in the majority of the exconjugants the transposon was located in the chromosomal DNA. Only in one case, exconjugant 2 of _Enterobacter agglomerans_ 339, Tn_5_ was plasmid-located.

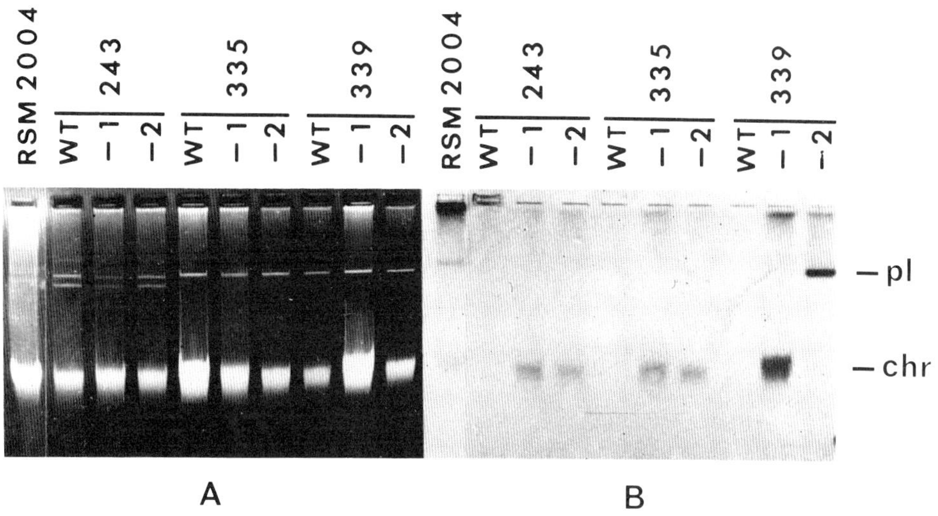

Figure 1. Agarose gel separation (A) and hybridization properties (B) of DNA from several exconjugants. Crude lysates from strain RSM2004 (donor), and Enterobacter agglomerans 243, 335 and 339, either wild type (WT) or exconjugants after mating (-1 and -2: two different exconjugants each). Hybridization was performed with biotinlabeled Tn5. chr, chromosomal DNA; pl, plasmid DNAs. For other details see text.

Survival of strain RSM2004 in pots with soil

The Enterobacter strains used in the above experiment had been isolated from fields around Bayreuth. Our detection of Tn5 transfer from RSM2004 to some of these strains on plates in the laboratory prompted us to check for conjugation in soil samples. We filled soil of the experimental field into pots and inoculated with strain RSM2004 as donor and Enterobacter agglomerans 339 as recipient, respectively. The experiment was carried out in the growth chamber. Samples were taken in suitable intervals to determine the colony forming units (cfu) of donor, recipient and exconjugants. We did not detect any kanamycin resistant exconjugants in this experiment. Our data document a marked decrease in survival, in particular for the recipient strain (Figure 2).

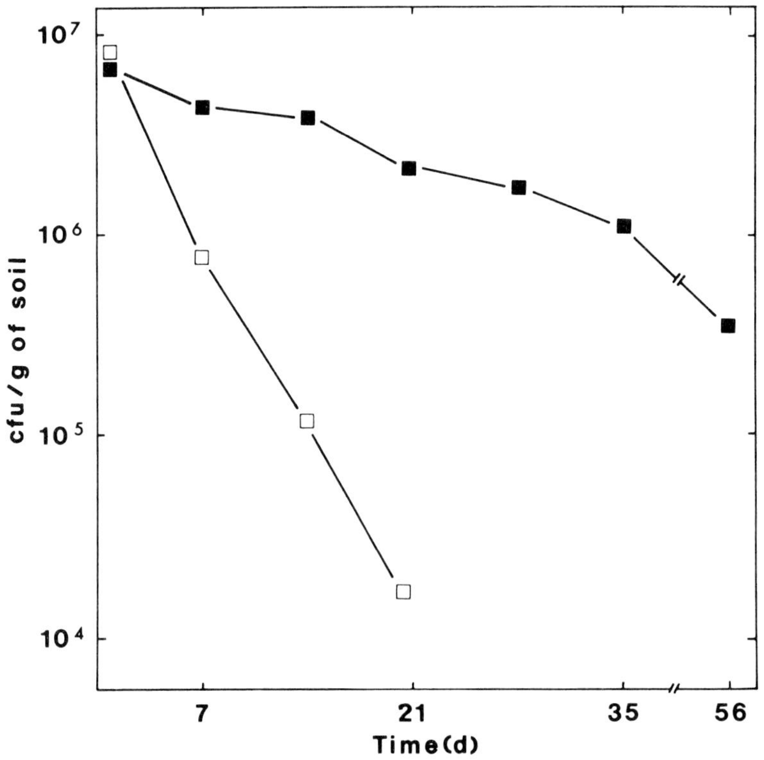

Figure 2. Survival of strains R. leguminosarum bv. viceae RSM2004 (■-■) and Enterobacter agglomerans 339 (□-□) in pots with soil from the experimental plot. For details see text.

Releasing experiment

The conditions for the release of R. leguminosarum bv. viceae RSM2004 and the planting diagram of the experimental plot at Bayreuth were almost the same as described by Hirsch and Spokes (3). Soil sampling for determining survival of the released strain was done from 12 sites distributed over the whole area of 200 m^2 inoculated with strain RSM2004.

To examine the nodulation ability of the released strain in the environment, pea plants grown at the inoculated field were carefully dug out, the nodules were sterilized and crushed. The cells within the nodules were then checked for their ability to grow on selective media. Nodules which contained cells growing on rifampicin and kanamycin and showed hybridization with Tn5 were formed by the released strain RSM2004. In that way we examined nodules of pea plants taken 51, 74 and 92 days after sowing seeds and we found that the quantity of nodules harboring strain RSM2004 relative to the total number of nodules decreased from 8 to 2 per cent (Table 2).

Table 2. Quantity of nodules from peas grown on experimental field that contain cells of *R. leguminosarum* bv. *viceae* RSM2004

Days after pea sowing	Number of nodules examined	%*
51	122	8
74	184	5
92	120	2

* : Quantity (%) of nodules with *R. leguminosarum* bv. *viceae* RSM2004 relative to the total number of nodules

Survival of RSM2004 in the inoculated field was investigated by two methods. The first one was dilution of soil samples from the field in buffer and direct spreading of aliquots on selective YM agar plates with rifampicin and kanamycin. This method, however - due to indigenous bacteria- is not very sensitive, not selective and showed high variability. Nevertheless the data allow a preliminary statement about the approximate spreading and the survival of the released strain. The results depicted in Figure 3 show a rapid decrease of the number of the colony forming units (cfu) of strain RSM2004 after release into the field both the pea and the cereal. The later increase of cfu in the pea-field corresponds with senescence and breaking up of the nodules formed by strain RSM2004.

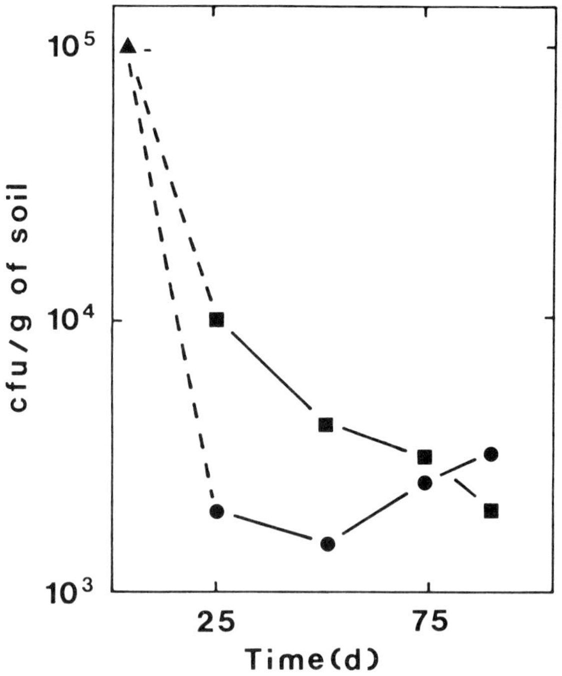

Figure 3. Survival of the released strain R. leguminosarum bv. viceae RSM2004 in the field, plotted against time. The first mark (▲) is a theoretical value obtained by considering the actual number of released cells of RSM2004 per g of soil. ■ - ■, cereal-field; ● - ●, pea-field.

The second, more sensitive and selective method to determine the survival of strain RSM2004 in the environment is to inoculate sterilized pea seeds in the laboratory with soil extract and grow the plants until they have formed nodules. Due to the high specifity of peas for R. leguminosarum bv. viceae the number of such bacteria in the soil can be derived by counting the nodules formed, referring to the Fisher and Yates tables (6). First results achieved with this method seem to confirm the results obtained by the direct plate method.

Discussion

The reported data, still at the beginning of the time schedule of our project, which is planned over three years, do not permit any final conclusions about gene transfer in the environment at this moment. We have established that transfer of Tn5 to other bacteria than Rhizobiaceae via plasmid pRL1JI is possible in the laboratory, but the transfer rate of kanamycin resistance is very low. We did not find stable replication of the plasmid in the exconjugants, but gene transfer occured by transposition of Tn5 to the recipient DNA. Conjugation experiments in pots of soil did not yield exconjugants, although the concentration of RSM2004 administered as donor and of the Enterobacter strain offered as recipient was about ten to hundred times higher than the concentration of the released strain in the experimental field. Hence gene transfer from the inoculated strain RSM2004 to nonrhizobia in the field would seem to be unlikely. However, the local concentration of the released strain RSM2004 could be very high at the moment of spreading the peat-based granular inoculant along the furrows before planting the seeds. Later, after a period of a few weeks, the released strain was spread and there was no difference in the number of cells of strain RSM2004 in the furrows and between the furrows. Nevertheless it seems crucial to improve the isolation and selection methods to detect even minute numbers of possible hybrid exconjugants. Our results show a rapid decrease of the titer of the released strain, and the nodulation ability of this strain in the field was less than expected from growth chamber experiments, again indicating that the released bacteria did not survive well in the native environment. The field site where the marked strain was released will be monitored for the next two years by planting various leguminous plants and studying their nodules. If gene transfer occurs, we should be able to isolate nodules yielding Tn5 containing cells.

Acknowledgements

We thank Marion Steinlein and Christine Fentner for excellent technical assistance and Penny Hirsch for providing the inoculant and the seeds. The help of John Spokes on sowing the experimental field is greatly appreciated. This work was supported by a grant from the EEC Biotechnology Action Programme in Risk Assessment.

References

1. Johnston A.W.B., Beynon, J.L., Buchanan-Wolloston, A.V., Setchell, S.M., Hirsch, P.R. and Beringer, J.E. (1978). Nature 276, 634-636.
2. Hirsch, P.R. (1979). J. Gen. Microbiol. 113, 219-228.
3. Hirsch, P.R. and Spokes, J.R. (1988). (This volume).
4. Simon, R., Priefer, U. and Pühler, A. (1983). Bio/technology 1, 784-791.
5. Maniatis, T., Fritsch, C.F. and Sambrook, J. (1982). Molecular cloning. A laboratory manual. Cold Spring Harbour Laboratory, USA.
6. Vincent, J.M. (1970). A manual for the practical study of root-nodule bacteria. Blackwell Scientific Publications, Oxford, UK.
7. Klingmüller, W. (1984). In: Trends in Molecular Genetics (Eds. Sinha, U. and Klingmüller, W.), pp. 37-61, Spectrum Publishing House, Delhi.
8. Kado, C.I. and Liu, S.T. (1981). J. Bacteriol. 145, 1365-1373.
9. Birnboim, H.C. and Doly, J. (1979). Nucl. Acid. Res. 7, 1513-1523.
10. Bethesda Research Laboratory Manual 'BluGene™: Nonradioactive Nucleic Acid Detection System'.
11. Leary, J.J., Brigati, D.J. and Ward, D.C. (1983). Proc. Natl. Acad. Sci. USA 80, 4045-4049.
12. Kleeberger, A., Castorph, H. and Klingmüller, W. (1983). Arch. Microbiol. 136, 306-311.
13. Singh, M., Kleeberger, A. and Klingmüller W. (1983). Mol. Gen. Genet. 190, 373-378.
14. Boyer, H.W. and Roullant-Bussoix, D. (1969). J. Mol. Biol. 41, 459-472.
15. Appleyard, R.K. (1954). Genet. 39, 440.
16. Kahn, M. and Helinski, D.R. (1978). Proc. Natl. Acad. Sci. USA 75, 2200-2204.
17. Dixon, R., Kennedy, C., Kondorosi, A., Krishnapillai, V. and Merrick, M. (1977). Mol. Gen. Genet. 157, 189-198.
18. Moore, R.J. and Krishnapillai, V. (1982). J. Bacteriol. 149, 276-283.
19. Krishnapillai, V., Nash, J. and Lanka, E. (1984). Plasmid 12, 170-180.
20. Döhler, K., Huss, V.A.R. and Zumft, W.G. (1987). Int. J. Syst. Bacteriol. 37, 1-3.
21. German Collection of Microorganisms. Mascheroder Weg 1b, D-3300 Braunschweig, FRG.

INTRODUCTION OF RHIZOBIUM INTO SOILS

Noëlle Amarger
Laboratoire de Microbiologie des Sols, INRA, BV1540
21034 Dijon Cedex, France.

Summary : Preliminary results of the common experiment designed to investigate gene transfer in the environment between introduced *Rhizobium leguminosarum* bv *viceae* RSM2004 and native strains of bacteria are reported. They show an exclusive nodulation of the pea by wild type strains, which casts some doubt on the successful introduction of the strain. General considerations on the introduction of Rhizobium into soils are given.

Keywords : Rhizobium, inoculation, competition.

Introduction

Since Rhizobium has been described as being responsible for producing a dinitrogen fixing symbiosis with legumes, it has been deliberatly released into the soil through seed inoculation. Although being more or less successful in increasing nitrogen fixation by legumes, inoculation with Rhizobium is still widely used all over the world. If we consider that 1 or 2 x 10^{7} hectares are inoculated each year with 10^{10} to 10^{11} rhizobia per hectare, we can see that at least 10^{17} rhizobia have been released each year for many decades in very diverse environments. From this pool of newly introduced genes interacting with the many species forming the indigenous populations, no rearrangement conducting to a species harmful to plants or man has been reported yet.

Experiments aimed at predicting the fate of genetically engineered bacteria into natural environment are conducted in different countries using soil samples in greenhouse experiments. However it is not known how predictive of what is going to happen in nature these experiments can be. As more knowledge will be gained from this type of experiments, it will appear necessary to go further, that is to say to go to the field. It seems logical to think, based on what was said above, that Rhizobium will be

Risk Assessment for Deliberate Releases
Edited by W. Klingmüller
© Springer-Verlag Berlin Heidelberg 1988

accepted as the least bad choice to make the first step. If Rhizobium is to be used as a model bacterium for genetic transfer experiments in soils, it is important to know how to introduce it. The problems encountered in the introduction of Rhizobium in soils will be briefly discussed after having reported preliminary results of the field experiment made in collaboration with P. HIRSCH and W. KLINGMULLER.

Release of genetically marked **Rhizobium leguminosarum** bv **viceae** RSM 2004 in field.

The aim of the experiment, the construction of the strain and the experimental design have been described in the previous paper by P. HIRSCH. In France the field was located at the INRA experimental farm near Dijon. The soil is a heavy clay which was found to contain between 10^4 and 10^5 R. leguminosarum bv viceae per g of dried soil using the M.P.N method (1). The experimental plot was 12 x 24 m, half was sown with peas, cv Solara, and the other half with spring wheat, on March 13 1987. Both cultures were inoculated at sowing using a special seeder which deliver liquid inoculant through a peristaltic pump on the seeds before closing the furrow. The liquid inoculant contained 2.8×10^8 R. leguminosarum bv viceae RSM 2004 per ml and 22 ml were delivered per m^2 which provided 6×10^9 bacteria per m^2.

Nodules from peas were sampled at the beginning of June and 2 weeks later. After surface sterilization their content was checked for resistance to kanamycine (40 ug ml^{-1}) streptomycine (100 ug ml^{-1}) and rifampicine (50 ug ml^{-1}). None of the nodules checked appeared to contain RSM 2004. The content of some nodules grew slightly on the kanamycin containing medium. The level of kanamycine resistance of these isolates was lower than the level of resistance usually given by Tn 5 and none of these isolates grew on a medium containing 20 ug per ml of streptomycine, level of resistance usually acquired by Tn 5 containing R. leguminosarum. This suggests that these isolates were not Tn 5 containing strain but wild type kanamycine resistant strains. To confirm this, hybridization with Tn 5 probe will be made.

It is difficult to say from these results if the non recovery of the inoculant strain from the nodules was due to a poor ability of RSM 2004 to compete with the indigenous strains or if it was due to its poor survival in soil prior nodulation.

These results underline once more the difficulty which is encountered when trying to introduce new strains of rhizobia in soils containing indigenous rhizobia of the same species, and to extrapolate results from pots to field and from one soil to another one. I will just try now to point out which factors are to be considered when a Rhizobium strain is to be introduced.

Methods used to detect Rhizobia

The major difficulty when studying Rhizobium in natural environment is that there is no direct method available for the detection and enumeration of Rhizobia in the soil, except immunofluorescence (2,3) a technique limited to the case where antibodies against the inoculant strain are specific enough not to react with the many strains composing a soil population.

Therefore indirect methods based on the property of rhizobia to nodulate legumes have to be used to detect and enumerate rhizobia in soils (1). These methods are suitable to follow the evolution of a population nodulating a given host, or the introduction of a new species, but they do not allow to track a particular strain among other strains specific for the same host as the plant will make nodules with all the strains. In this case the inoculant strain must have some characteristic different from the soil strains, which can be easily identified, in order to distinguish between nodules made by the inoculant strain from the others. Many different techniques have been utilized for this purpose : serology with all its derivatives (4, 5, 6), bacteriophage (7) or intrinsic antibiotic resistance (8, 9), antibiotic resistant mutants (12, 13), sodium dodecyl sulfate polyacrylamide gel electrophoresis (10, 11). The simplest technique which easily allows screening of high numbers of nodules is the use of antibiotic resistance mutants. However mutations to resistance to antibiotics can change the symbiotic properties of the Rhizobium strains (14, 15).

Identification of the strains at the origin of nodules is not however sufficient to detect and enumerate the strains which are in the soils, as a legume in presence of different strains may favor one strain to form nodules. Rhizobium strains differ in their capacity to be selected by the plant host, that is to say in their nodulating competitiveness. Consequently a non competitive strain, although present in a soil, may well not be detected by the only methods available and enumeration of Rhizobium strains belonging to a same inoculation group is therefore completely impossible.

Factors contributing to Rhizobium competition

The numbers of nodules made by a strain inoculated to a legume planted in a soil containing indigenous rhizobia of the same specificity, will depend not only on the differences of nodulating competitiveness between inoculant strain and soil strains but also on the relative number of each type present at the sites of infection. This number will itself be dependent on the number of the inoculant strain which survived inoculation and competition with other soil microorganisms before it reached the infection sites. The competition between inoculant and indigenous strains is therefore dependent on three main variables : the number of inoculant bacteria as compared to the number of soil strains, the relative ability of inoculant and soil strains to survive before nodulation, a step which has been called saprophytic competence (16), and the relative capacity of inoculant and soil strains to be selected by the host plant.

Survival of rhizobia in soil and rhizosphere before nodulation is related to intrinsic properties of each strain. It will depend on how each strain responds to different abiotic factors as temperature, pH, dessication, nutrients content and form, salinity and how they withstand parasitism, predation and the action of microbial antagonists. The results of this first level of competition which are numbers of inoculant and soil bacteria cannot be assessed directly as we have seen that there is no mean of counting different strains belonging to a same species.

The reasons why a given host will select one strain preferentially to another one are not known, but as different host varieties may differ in their choice host varieties x Rhizobium strains interactions are observed. This second level of competition can not be assessed either, as we do not the proportion of inoculant strain being present at the nodulation sites.

Measurement of the proportion of nodules made by the inoculant strain will therefore estimate the overall phenomenon of competition at a given level of inoculant bacteria. If the competitive ability of different inoculant strains are to be compared with a same soil, they need to be inoculated at an identical level. If the competitive ability of one strain is to be compared with different soils, this will be possible only if the numbers of indigenous bacteria are the same in the soil, unless the relationship between proportion of competing bacteria and proportion of nodules are known (17). When looking at the proportion of nodules made by a same inoculant strain in different soils, using such a relationship, it was found (18) that soils had a high effect on the success of inoculant strain to nodulate its host. Two hundred times more inoculant bacteria could be necessary in one soil compared to another one to produce a same number of nodules, indicating big differences in the soils competitiveness. When two strains of different competitiveness were compared in two soils, the difference in competitiveness was not the same indicating a strain x soil interaction. This soil effect could be due either to its effect on the survival of inoculant bacteria or could be due to the nodulating competitivity level of the indigenous strains. The soil x strain interaction would indicate that the niche available varies with strains and with soils, some strains being more adapted to some soils.

It is interesting to remark that in the absence of competing rhizobia the introduction of Rhizobium is realized without any problem. For example rhizobia nodulating soybean are absent in soils from European countries. From a 10 years experience in France we know that once inoculated soybean has grown in a field, the soil contains high numbers, between 10^3 and 10^5 per g, of the inoculant strain. The strain persists at the same level, for at least the time of our experience, even in the absence of a new soybean crop. The same observation can be made for the rhizobia nodulating lupine which are absent in neutral or alkaline soils, but can be found at a level of 10^3, 10^4 per g. in the same soils after inoculated lupine has been grown. It seems that there is no problem to introduce a Rhizobium species which is not already present in soils, at least in temperate conditions and fertile soils. We must admit that there are some ecological niches available for specific rhizobia which are not yet represented in soils.

Conclusion

It can be considered that in each soil, a given species of Rhizobium has at its disposal a special niche, the size of which will vary with the soil but which is usually big enough to accomodate 10^3 to 10^5 bacteria per g. When this specific niche is empty there are no problems in having it occupied by any strain of the right specificity. But when the niche is already occupied, to replace the occupants necessitates to give them some advantages which can be of three different types, number, better adaptability to soils and high selectivity by the plant. In our field experiment to improve introduction of the inoculant strain, the number of inoculant bacteria can only be raised by a factor of near five but not more, nothing can be done to better adapt the organism to soil environment as we have no idea of what is limiting, and we can try to select a strain with high capacity to be selected by the host and genetic markers which do not alter their symbiotic properties, which may be difficult to find. Another possibility might be to create a new niche by utilizing a new host, afghan peas, which require R. leguminosarum bv viceae of enlarged specificity, not present in european soils. It has to be checked first if such a new niche is created or if these rhizobia have to share the common bv viceae niche.

References

1. Vincent, J.M. (1970). A manual for the Practical study of Root-Nodule Bacteria. Blackwell Scientific Publications, Oxford, UK.
2. Bohlool, B.B. and Schmidt E.L. (1980). Adv. Micro. Ecol. 4, 203-235.
3. Bohlool, B.B., Kosslack, R. and Woolfenden, R. (1984). In Advances in Nitrogen Fixation Research (Eds Veeger C. and Newton W.E.) p 287-293.
4. Moawad H.A., Ellis W.R. and Schmidt E.L. (1984) Appl.Environ. Microbiol. 47, 607-617.
5. Berger J.A., May S.N., Berger L.R. and Bohlool B.B. (1979) Appl. Environ. Microbiol. 37, 642-646.
6. Kishinevsky B. and Maoz A. (1983) Current Microbiol. 9, 45-49.
7. Lesley, S.M. (1982) Can. J. Microbiol. 28, 180-186.
8. Benon, J.L. and Josey, D.P. (1980). J. Gen. Microbiol. 118, 437-443.
9. Jenkins, M.B. and Bottomley P.J. (1985) Soil Biol. Biochem. 17, 173-179.
10. Noel, K.D. and Brill, W.J. (1980) Appl. Environ. Microbiol. 40, 931-938.
11. Fuquay, J.I., Bottomley P.J. and Jenkins M.B. (1984) Appl. Environ. Microbiol. 47, 663-669.

12. Kuykendall, L.D. and Weber D.F. (1978) Appl. Environ. Microbiol. 36, 915-919.

13. Amarger N. (1974). C.R.A.S. Paris 279, 527-530.

14. Lewis, O.M., Bromfield, E.S.P. and Barran, L.R. (1987) Can. J. Microbiol. 33, 343-345.

15. Amarger N. (1975) C.R.A.S. Paris 280, 1911-1914.

16. Chatel, D.L., Greenwood, R.M. and Parker, C.A. (1968) In Transactions of the Ninth International Congress of Soil Science Adelaïde Australia. p.65-73.

17. Amarger, N. and Lobreau, J.P. (1982). Appl. Environ. Microbiol. 44, 583-588.

18. Amarger N. (1984) In Current Perspectives in Microbial Ecology. American Society for Microbiology, Washington, p. 300-304.

STRATEGY FOR MONITORING PEA-NODULATING RHIZOBIA WITHOUT IN VITRO GENE MANIPULATION

Wolfgang Lotz
Institut für Mikrobiologie und Biochemie
Lehrstuhl für Mikrobiologie, Universität Erlangen-Nürnberg
Staudtstr. 5, D-8520 Erlangen, FRG

Summary: A strategy is described for monitoring pea-nodulating bacteria of *Rhizobium leguminosarum* B10 (Nod^+, Fix^+, Hup^+) without prior *in vitro* gene manipulation. The rhizobia are released into the soil of a selected test plot (having a low level of resident Hup^+ *R.leguminosarum* bacteria) via sowing of B10-inoculated pea seeds. Seeds of the original host of strain B10, *Pisum sativum* var. Poneka, are used for the release experiment and for further nodulation assays. Rhizobia isolated from sampled root nodules of pea plants grown in the test plot will be screened for the following properties: pattern of antibiotic resistance, presence of *hup*-specific DNA, UV-inducible bactericidal agent, plasmid pattern.
Keywords: *Rhizobium leguminosarum*, *Pisum sativum*, *hup* genes, plasmids, seed inoculation, monitoring strategy.

I. Introduction

The enquete commission on "Chancen und Risiken der Gentechnologie" of the 10th German Bundestag has recently recommended a moratorium of five years on the release of microorganisms genetically engineered by *in vitro* manipulation (1). In order to assess the risks possibly created by the mass-release of microorganisms into the environment, specific monitoring methods have to be developed using a combination of genetical and ecological strategies. In the model study described here, specific sequences of cloned wild type DNA of *Rhizobium leguminosarum* B10 are used as hybridization probes for the monitoring of released bacteria of this strain. *R.leguminosarum* B10 can nodulate (Nod^+), fix nitrogen (Fix^+), and oxidize hydrogen (Hup^+) in symbiosis with pea plants (2). The Hup^+ phenotype, which is mediated by an uptake hydrogenase, has been found relatively rarely among bacteria of the genus *Rhizobium* (3). It is planned to release bacteria of the Hup^+ strain B10 into a test plot containing mainly Hup^- resident (pea nodulating) *R.leguminosarum* strains.

Risk Assessment for Deliberate Releases
Edited by W. Klingmüller
© Springer-Verlag Berlin Heidelberg 1988

II. Site-specific occurence of Hup$^+$ Rhizobium leguminosarum strains

Among R.leguminosarum strains isolated from pea root nodules, site-specific differences in the occurence of Hup$^+$ strains have been demonstrated (4). These differences seem to be stabley maintained over many years. A total of 265 R.leguminosarum strains has been isolated during the years 1980, 1982, 1983, and 1984 from the root nodules of Pisum sativum var. Poneka grown at two locations near Erlangen. Location "B" is a garden bed in which pea plants have been grown annually since 1975, and "W" is a meadow where pea plants have not been grown for many years. For the isolation of rhizobial strains from location "W" only a small number of pea plants were grown at different sites. Only one strain has been isolated per nodule; 135 strains were from "B" and 130 strains from "W". All strains were Nod$^+$, Fix$^+$ with P.sativum var. Poneka. About 70% of the strains isolated annually from location "B" were Hup$^+$, but only two of the W-strains showed a Hup$^+$ phenotype. The Hup-phenotype has been determined with a methylene blue reduction assay (5).

III. Construction of a hup-specific DNA probe

R.leguminosarum B10 has been isolated from location "B" in 1980 (6, 7). A cosmid gene bank of its total DNA in vector pMMB34 has been constructed (2). Clone pR1B505 from this bank carries DNA sequences homologous to the hup-specific clone pHU1 of Bradyrhizobium japonicum (8). All EcoRI-fragments of the pHU1 insert, except the smallest one, hybridized to the insert (BamHI-2 fragment) of pR1B505. Restriction enzyme cleavage sites of pR1B505 have been mapped (2, 9). A region of 4,4 kb of the HindIII-2 fragment (10,2 kb) of pR1B505 hybridized strongly to the HindIII-1 fragment of pHU1, which codes for the 60.000 kDa subunit of the uptake hydrogenase of B.japonicum (10). The HindIII-2 of pR1B505 may therefore carry a structural gene for the uptake hydrogenase of R.leguminosarum B10. The cosmid clone pR1B505 has been subcloned into pACYC184 (Fig. 1). Subclone pHVT116 carrying the HindIII-2 fragment has been used as a probe for the detection of hup-specific sequences in newly isolated Rhizobium leguminosarum strains. In addition, the small SalI-6 fragment of pR1B505, located on the HindIII-2 fragment and subclone pHVT109 (HindIII-5 fragment of pR1B505) have also been used as probes in DNA/DNA-hybridization experiments.

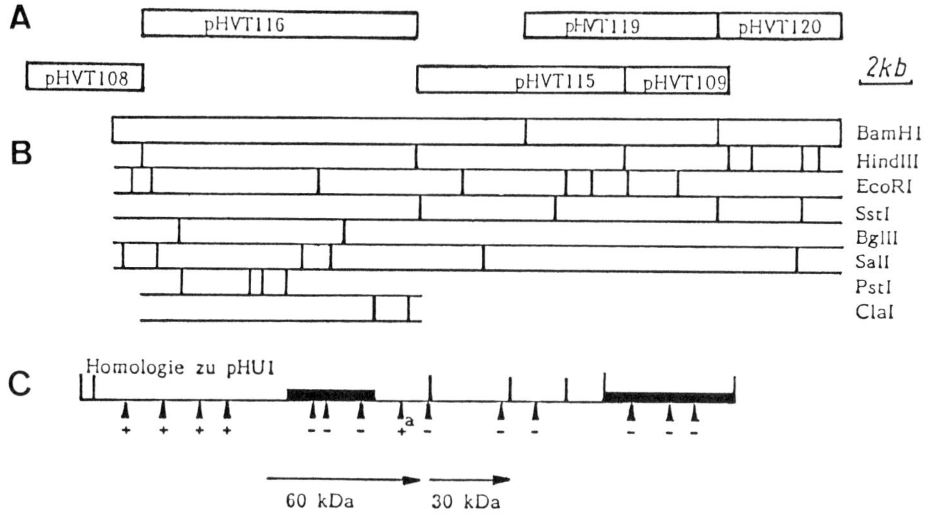

Figure 1. (A) Subclones of pR1B505. (B) Restriction map of the insert of pR1B505 carrying DNA of the Sym-plasmid of R.leguminosarum B10. (C) EcoRI restriction map of cosmid pHU1 carrying hup-specific DNA of B.japonicum (8). The Hup phenotype of Tn5-insertion mutants and the probable location of the structural genes for the 60 kDa and the 30 kDa subunits of the uptake hydrogenase of B.japonicum have been determined (10), (a = weakly positive Hup-phenotype). ▬: homology between inserts of pR1B505 and pHU1 (2, 9).

IV. Criteria for the screening assay

Isolation of pea root nodule bacteria

Field trials provide the ultimate test of strain performance (e.g. persistence in the soil and competition for nodulation), since the proportion of nodules formed by the released strain is the end result of the interactions of all factors involved (11). Pea plants are grown in the test plot in order to "sample" the population of rhizobial strains nodulating the particular "trap" host, P.sativum var. Poneka, from which R.leguminosarum B10 has been isolated. Only one bacterial strain is isolated per sampled, surface-sterilized root nodule. The isolated strains will be replica-plated for the colony-hybridization and the drug-resistance assays described below. With the mentioned strategy only a small and specific sector of the broad spectrum of soil microorganisms is selected. This approach does not allow the assessment of the proportion of released R.leguminosarum B10 bacteria present among the free-living R.leguminosarum strains in the test plot.

Screening for drug resistance

The drug-resistance pattern of strains isolated from the sampled root nodules will be compared with the B10-specific pattern. R.leguminosarum B10 shows an intrinsic resistance to penicillin, but is sensitive to other drugs such as rifampicin, streptomycin, tetracyclin, chloramphenicol, and kanamycin. In order to persist in competition with the normal microbiota of a given test plot, the inoculant strain should carry the drug resistance marker(s) most dominant among the (pea-nodulating) resident rhizobia tested. In addition, the strain should carry a drug-resistance marker ("screening marker") only rarely found among the local strains. Spontaneous antibiotic resistance mutants of a given rhizobial wild type strain can be obtained relatively easily on selective plates. Special care must be taken in choosing symbiotically effective strains among the mutants. In the case of R.leguminosarum, Pain (12) has found that of eight antibiotics tested, rifampicin was the only one for which resistant mutants were symbiotically defective. Similar results have been obtained with rifampicin resistant mutants of R.meliloti (13), indicating that the rif^R marker is unsuitable for ecological studies.

Screening for hup-specific DNA

The cloned hup DNA of R.leguminosarum B10 can be used as a probe in order to identify rhizobial strains carrying hup genes. The DNA/DNA hybridization experiments can be done via colony hybridization (14). This technique has been used in order to identify Rhizobium trifolii strains from the root nodules of clover, Trifolium subterraneum (15). Recently, a method has been described for the identification of Lotus rhizobia by direct DNA hybridization of crushed root nodules (16). Possibly, such an approach could be developed into a relatively fast, standardized assay for the screening of specific DNA-sequences in legume root nodules. When using a crushed nodule directly for the DNA/DNA-hybridization, special care must be taken to remove bacteria (e.g. of the inoculant strain) from its surface.

Recent studies indicate that the surface of hydrogen evolving legume root nodules may offer a niche for hydrogen oxidizing soil bacteria (17, 18). Wong et al. (18) have demonstrated an enrichment for hydrogen-oxidizing Acinetobacter spp. in the rhizosphere of hydrogen-evolving soybean root nodules.

Cloned hup-DNA from R.leguminosarum B10 has been used as a hybridization probe for the search of homologous sequences in Acinetobacter calcoaceticus BD4 (19) and in aerobic,hydrogen oxidizing Alcaligenes eutrophus (ATCC 17699). Using subclones PHVT109 and pHVT116 (Fig. 1) as probes, hybridization signals could not be detected with the total EcoRI-digested DNA of these strains. The hybridization conditions have been described (20). Under these conditions the probes pHVT109, pHVT116, and the SalI-6 fragment of pR1B505 did not hybridize to the EcoRI-digested insert DNA of plasmids pCH128 and pCH129 (hybridization temperature was either 35 °C or 42 °C). Plasmids pCH128 and pCH129 carry the genes for soluble and particulate hydrogenase of A.eutrophus, respectively (21, 22); they have been kindly provided by B. Friedrich (Institut für Pflanzenphysiologie, Zellbiologie und Mikrobiologie, Freie Universität Berlin). Both pHVT109 and pHVT116 carry sequences strongly hybridizing with the Bradyrhizobium japonicum-specific DNA of pHU1 (Fig. 1). Obviously, homology between the tested sequences of strain B10 and B.japonicum must be significantly higher than between the cloned B10-DNA and the tested sequences of A.eutrophus H16 or A.calcoaceticus BD4. In this context, it is interesting that antibody raised against B.japonicum hydrogenase cross-reacted with purified hydrogenases of four hydrogen-oxidizing bacteria to various degrees and in the following order: B.japonicum > Alcaligenes latus > Alcaligenes eutrophus > Azotobacter vinelandii (23). Antibody to B.japonicum hydrogenase, inhibiting the H_2 oxidation activity of the respective enzymes, followed the same pattern; no inhibition of A.vinelandii hydrogenase activity was detected.

Screening for phage production

R.leguminosarum B10 has been shown to produce a UV_{260}-inducible bactericidal agent with a specific host range (6). This agent has been identified as bacteriophage WK of strain B10 (24). Due to spontaneous and UV-induced phage production inhibition zones are formed around spots or colonies of strain B10 when a WK-sensitive indicator strain is used (Fig. 2). The specific host-range has also been demonstrated for phage WK (24). Therefore, the formation of the inhibition zones can be used as a criterium in screening for strain B10.

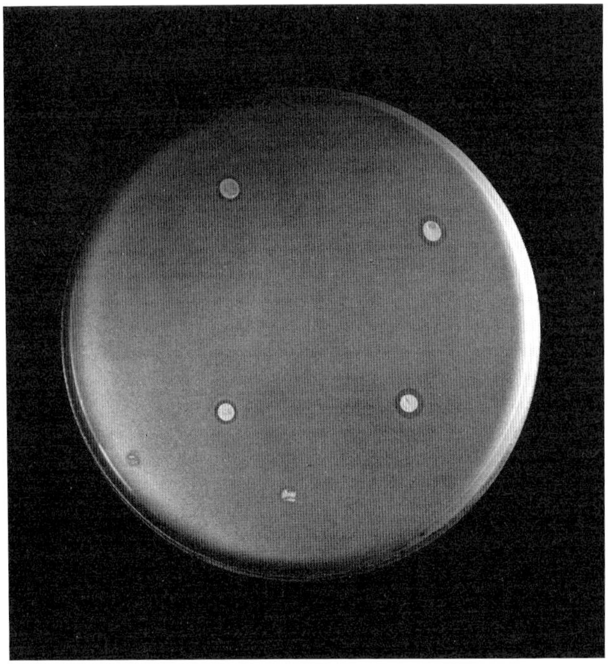

Figure 2. Production of inhibition zones around white spots of *Rhizobium leguminosarum* B10 growth. The inhibition zones are formed by bacteriophage WK in a lawn of WK-sensitive *R.leguminosarum* B6 indicator bacteria. Like strain B10, strain B6 has been isolated from a (separate) root nodule of *Pisum sativum* var. Poneka grown at location "B" in 1980. Only spots on the right half of the agar plate have been UV_{260}-irradiated. After overnight incubation at 25 °C, all cells were inactivated by chloroform vapor. After evaporation of the chloroform, the plate was overlayered with soft-agar containing cells of the WK-sensitive indicator strain B6. Growth of the cells was on TY-agar (25). Foto by Tichy (6).

Analysis of the rhizobial plasmid pattern

With all Hup^+ *R.leguminosarum* strains isolated from the root nodules of pea plants grown at locations "B" and "W", *hup*-specific sequences were always occuring together with *nod*- and *nif*-specific determinants on the Sym(biosis)-plasmid of a given strain (4, 26). Similar results have been obtained for other *R.leguminosarum* strains (20, 27). Except for two strains, the largest plasmid (>500 Mdal) per B- or W-strain was not a Sym-plasmid. The molecular weights of most Sym-plasmids were in the range of 170 to 350 Mdal. Two to six plasmid-size classes were found per strain. A wide variety of plasmid patterns was obtained per location and year of isolation. Plasmid patterns most common per year and location occured with frequencies of about 25% to 50%. For B- and W-strains respectively, different patterns dominated each year of isolation. In R.leguminosarum B10 four plasmid-size classes have been detected by agarose gel electrophoresis: >500 Mdal ("megaplasmid"), ~300 Mdal (Sym-plasmid), 150 Mdal, and 17 Mdal (6, 7).

V. Program of strain release and monitoring strategy

Selection of test plot

The release experiment is planned for the growing season of 1988. Before the mass-release of R.leguminosarum B10 bacteria, the test plot will be chosen among a number of candidate plots according to the following criteria:

(1) Low level of pea-nodulating Hup^{+} R.leguminosarum strains (e.g. location W).

(2) Difference in drug resistance patterns of the majority of resident nodule bacteria and the inoculant strain B10.

(3) Previous history of the candidate plot (e.g. crop rotation, application of fertilizer and agrichemicals).

The general design of such field experiments has been described (11).

Release of R.leguminosarum B10

R.leguminosarum B10 will be mass-released as pea seed inoculant. Seeds of the original host of strain B10, P.sativum var. Poneka, will be used in the assay. A major advantage enjoyed by the inoculant bacteria is their proximity to the developing rhizosphere of the potential host plant. Before the release, a fresh culture (10^{9} cells/ml) of strain B10 is mixed with autoclaved, sterile peat. After the addition of gum arabicum as an adhesive, the pea seeds are mixed with the slurry and sowed into the soil. This procedure protects the in-occulant bacteria against possible deleterious environmental factors. The preparation of the legume inoculant (11) and its application (28, 29) have been described.

Monitoring the released rhizobia

After the release of strain B10 via sowing of the inoculated pea seeds, the screening assay described (see IV.) will be used for monitoring. The assay includes the following steps:

(1) Sampling of pea root nodules six to eight weeks after sowing.
(2) Screening the nodule bacteria for the drug resistance pattern.
(3) Screening for the presence of hup-specific DNA in the nodule bacteria.
(4) Screening for UV_{260}-inducible bactericidal agent (WK-type host range).
(5) Screening for B10-type plasmid pattern.

Since B10-inoculated pea seeds were used for the release experiment, a significant portion of the root nodules of plants obtained from these seeds can be expected to contain B10 bacteria. In all further screening assays, surface-sterilized pea seeds (again of variety Poneka) will be sowed in the test plot, in order to "trap" R.leguminosarum strains from the soil. One or two screening assays per year over a period of three years are planned. During this long-time study, the history of the test plot (e.g. crop rotation) will be documented.

Perspective

Stacey (30) has pointed out in an excellent review on the ecological and genetical aspects of rhizobial strain release: "Although only limited experimental data are available, it is possible that transfer of at least plasmid-encoded traits is common among Rhizobium strains in the soil. Such transfer would most likely occur in the rhizosphere or in the nodule, where cell numbers are elevated. Such gene transfer would create new gene mixtures which could be selected for or against by the interaction of the myriad of biotic and abiotic factors". It was further pointed out: "Genetic persistence of environmentally introduced foreign traits should be considered separately from organism persistence". In order to study the ecology of "released genes" of Rhizobium leguminosarum, further hybridization probes and assays have to be developed.

Acknowledgements

This work was financially supported by the Deutsche Forschungsgemeinschaft and the Bundesministerium für Forschung und Technologie.

References

1. Chancen und Risiken der Gentechnologie, Enquete-Kom. d. 10. Dt. Bundestages (1987). (Ed. Deutscher Bundestag, Referat Öffentlichkeitsarbeit), Bonn.
2. Tichy, H.V. (1986). Klonierung und Analyse von Aufnahmehydrogenase (hup) Genen aus Rhizobium leguminosarum B10. Universität Erlangen-Nürnberg, Dissertation.
3. Eisbrenner, G. and Evans, H.J. (1983). Ann. Rev. Plant Physiol. 34, 105-136.
4. Fees, H., Tichy, H.V. and Lotz, W. (1985). In: Nitrogen Fixation Research Progress (Eds. Evans, H.J., Bottomley, P.J. and Newton, W.E), p. 412, Martinus Nijhoff Publishers, Dordrecht.
5. Tichy, H.V. and Lotz, W. (1985). FEMS Microbiol. Lett. 27, 107-109.
6. Tichy, H.V. (1981). Charakterisierung der Plasmide neu isolierter Rhizobium leguminosarum Stämme. Universität Erlangen-Nürnberg, Diplomarbeit.
7. Tichy, H.V, and Lotz, W. (1981). FEMS Microbiol. Lett. 10, 203-207.
8. Cantrell, M.A., Haugland, R.A. and Evans, H.J. (1983). Proc. Natl. Acad. Sci. USA 80, 181-185.
9. Tichy, H.V., Schild, C., Ripke, H.M., Nelson, L.M., Fees, H. and Lotz, W. (1987). In: Molecular Genetics of Plant-Microbe Interactions (Ed. Verma, D.P.S. and Brisson, N.), pp. 279-281, Martinus Nijhoff Publishers, Dordrecht.
10. Zuber, M., Harker, A.R., Sultana, M.A. and Evans H.J. (1986). Proc. Natl. Acad. Sci. USA 83, 7668-7672.
11. Date, R.A. and Roughley, R.J. (1977). In: A treatise on Dinitrogen Fixation, IV. Agronomy and Ecology (Eds. Hardy R.W.F. and Gibson, A.H.), pp. 243-275, John Wiley & Sons, New York.
12. Pain, A.N. (1979). J. Appl. Bacteriol. 47, 53-64.
13. Lewis, D.M., Bromfield, E.S.P. and Barran, L.R. (1987). Can. J. Microbiol. 33, 343-345.
14. Grunstein, M. and Hogness, D.S. (1975). Proc. Natl. Acad. Sci. USA 72, 3961-3965.
15. Hodgson, A.L.M. and Roberts, W.P. (1983). J. Gen. Microbiol. 129, 207-212.

16. Cooper, J.E., Bjourson, A.J. and Thompson, J.K. (1987). Appl. Environ. Microbiol. 53, 1705-1707.
17. Cunningham, S.D., Kapulnik, Y. and Phillips, D.A. (1986). Appl. Environ. Microbiol. 52, 1091-1095.
18. Wong; T.-Y., Graham, L., O'Hara, E. and Maier, R.J. (1986). Appl. Environ. Microbiol. 52, 1008-1013.
19. Juni, E. and Janik, A. (1969). J. Bacteriol. 98, 281-288.
20. Nelson, L.M., Grosskopf, E., Tichy, H.V. and Lotz, W. (1985). FEMS Microbiol. Lett. 30, 53-58.
21. Eberz, G., Hogrefe, C., Kortlüke, C., Kamienski, A. and Friedrich, B. (1986). J. Bacteriol. 168, 636-641.
22. Friedrich, B., Kortlüke, C., Hogrefe, C., Eberz, G., Silber, B. and Warrelmann, J. (1986). Biochimie 68, 133-145.
23. Arp, D.J., McCollum, L.C. and Seefeldt, L.C. (1985). J. Bacteriol. 163, 15-20.
24. Kokotek, W. (1985). Isolierung und Charakterisierung eines bakteriziden Agens von Rhizobium leguminosarum B10. Universität Erlangen-Nürnberg, Diplomarbeit.
25. Beringer, J.E. (1974). J. Gen. Microbiol. 84, 188-198.
26. Seifert, B.L., Tichy, H.V., Nelson, L.M., Cantrell, M.A., Haugland, R.A. and Lotz, W. (1984). In: Advances in Nitrogen Fixation Research (Eds. Veeger, C., Newton, W.E.), p. 721, Martinus Nijhoff Publishers, The Hague.
27. Leyva, A., Palacios, J.M. and Ruiz-Argüeso, T.R. (1987). Appl. Environ. Microbiol. 53, 2539-2543.
28. Brockwell, J. (1977). In: A Treatise on Dinitrogen Fixation, IV. Agronomy and Ecology (Eds. Hardy, R.W.F. and Gibson, A.H.), pp. 277-309, John Wiley and Sons, New York.
29. Williams, P.M. (1984). In: Biological Nitrogen Fixation (Ed. M. Alexander), pp. 173-200, Plenum Press, New York and London.
30. Stacey, G. (1985). In: Engineered Organisms in the Environment: Scientific Issues (Eds. Halvorson, H.O., Pramer, D. and Rogul, M.), pp.109-121, American Society for Microbiology, Washington.

PRELIMINARY TRIALS OF FIELD RELEASE OF AZOSPIRILLUM BRASILENSE AS INOCULANT IN NORTHERN ITALY

Nuti M.P., Rubboli P.°

Dept. of Agricultural Biotechnology, University of Padova, via Gradenigo 6, 35131 Padova (Italy) - ° Società Europea del Seme, viale della Repubblica 19, 48024 Massa Lombarda (Italy)

Joint project on: Physiology and genetics of Azospirillum/plant interaction (partners: M. Bazzicalupo, Firenze; I. Cacciari, Roma; F. Favilli, Firenze; G. Morpurgo, Roma; G. Zanetti, Milano).

Summary

Two fields experiments were conducted with *Triticum durum* during 1985/86 and 1986/87 in northern Italy, to study inoculation effects of homologous and heterologous strains of *Azospirillum brasilense*.

The results were fairly fluctuating in the two years; in some cases a yield response to inoculation with the homologous strain was detectable, while Azospirillum strain Cd gave almost *nihil* response in all conditions tested. Soil organic matter content seems to play an important role, i.e. no response to inoculation has ever been obtained at 4%, some effects are just detectable at 3%, while they can be quantified in soils with 1.2-1.5% organic matter. The response to inoculation, when present, consisted in higher (up to 14%) yields, increased weight of root or aerial parts of the plant, higher concentration of Ca, Mg, Mn, Fe in roots or aerial parts of the plant. The persistence of inoculated Azospirilli in soil or

Risk Assessment for Deliberate Releases
Edited by W. Klingmüller
© Springer-Verlag Berlin Heidelberg 1988

rizosphere seems to be short, i.e. few weeks, and differed for the strain used.

Introduction

There are several reports on the inoculation of cereals with Azospirillum isolated from rizosphere soil or roots of heterologous plant species (1); however, increasing circumstantial evidence suggests that beneficial effects on the plant can be obtained, following inoculation with the homologous Azospirillum (4). Although the mechanism determining yield increase of inoculated plants remains to be completely elucidated, recent studies indicate a possible involvement of auxins and other plant-growth promoting substances (3). Genetic and physiological studies support this hypothesis (2).

In this comunication we report the results of two field experiments with Triticum durum conducted in 1985/86 (Exp. 1) and 1986/87 (Exp. 2) in northern Italy, to study inoculation effects of homologous (strain F14) and heterologous (strain Cd = ATCC 29710) Azospirillum brasilense.

Materials and methods

The wheat cultivar used was Creso, and Azospirillum strains were both rifampicin (40 μg ml^{-1}) spontaneous resistant mutants. Inocula were prepared on peat (powder or granular) and each plant received 1-5 x 10^{-5} cells.

Inoculated strains were recognized by their antibiotic marker and red pigmentation (strain Cd). Three levels of soil organic matter were compared in Exp. 1, i.e. 1.2, 2.9, and 4.0%; two levels were considered in Exp. 2, i.e. 1.5 and 3.1%. Plots were ca. 5000 m^2 each, and there were nine randomized complete blocks. Cultivation included pre-emergence and post-emergence weed control as usual in the area. Plants were harvested throughout the growth cycle; chemical analyses were performed on whole seedlings until one month after sowing, on aerial part until the grain harvest.

Results and discussion

Results from Exp. 1 show, in uninoculated plots, a significant yield response to N fertilization, the maximum being reached with 140 Kg x ha^{-1}; a yield decrease was constantly found with N application rate of 210 Kg x ha^{-1}; the decrease was particularly evident also in plots inoculated with homologous Azospirillum (Tab. 1).

Tab. 1 - Effect of inoculation with A.brasilense on yield (expressed in Kg x ha^{-1}) of Triticum durum cv. Creso, in a clay-soil containing 1.4% organic matter.

N fertilizers level (Kg x ha^{-1})	uninoculated control	inoculated with strain Cd	F14
0	3.030	3.530	3.140
70	5.240	4.910	6.100
140	6.480	6.510	6.800
210	6.300	6.150	6.240

A yield response, following inoculation, was only detected with strain F14 in plots receiving 70 Kg N x ha^{-1}. In Exp. 2 the response was only detected, with the same strain, but in plots receiving 140 Kg N x ha^{-1}. The reason for the observed fluctuation remains obscure. Organic matter content seems to play a key role: no response to inoculation has ever been obtained at 4%; some effects were just detectable at 3%, while the observed yields increased was obtained, for both experiments, in soils with 1.2-1.5% organic matter. Heterologous strain Cd gave almost nihil response in all condition tested; inoculation with strain F14 was followed by a slight increase of root weight, and later during the growth cycle by an increased weight of aerial parts. In the inoculated (F14) plots, at N = O and at N = 70 Kg x ha^{-1}, Ca and Mg were found to be higher in the root system, with respect to uninoculated controls; the same applies, in the shoot of inoculated (F14) plants, for Fe at any N fertilization level, while Ca, Mg, Mn, Zn and B tend to increase particularly with N = 70 Kg x ha^{-1}. MPN counts of Azospirilli showed that the presence in soil or rizosphere of the inoculated strain is short, i.e. few weeks, strain Cd persisted slightly more than F14. This would suggest that inoculation, when beneficial, should be carried out each sowing season. Overall, this research confirms and extends previous studies on the effects of cereal inoculation with Azospirillum, although the data presented here remain preliminary. We suggest that elements other than the usually considered N-P-K might be involved in the elicitation of yield response following inoculation with the appropriate strain.

Acknowledgments. The authors are grateful to Dr. F. Rosso (C.S.R.B., Ravenna), Dr. L. Pifanelli (Az. Le Gallare), Mr. L. Collina (Az. Imm. Dante) and Mr. M. Mancini (Az. Ca' Bosco) for providing skillful technical assistance and the facilities for field experiments. This work was supported in part by C.N.R., Special Grant IPRA (Paper no. 1522).

References

1. Baldani V.L.D., Baldani J.I. and Döbereiner J. (1987). Biol. Fert. Soils 4, 37-40.
2. Barbieri P., Zanelli T., Galli E. and Zanetti G. (1986). FEMS Microbiol. Lett. 36, 87-90.
3. Morgenstern E. and Okon Y. (1987). Arid Soil Res. Rehabil. 1, 115-127.
4. Warenburg F.R., Dreessen R., Vlassak K. and Lafont F. (1987). Biol. Fertil. Soils 4, 55-59.

THE DEVELOPMENT AND EXPLOITATION OF 'MARKER GENES' SUITABLE FOR RISK EVALUATION STUDIES ON THE RELEASE OF GENETICALLY ENGINEERED MICROORGANISMS IN SOIL.

Fergal O'Gara, B. Boesten and S. Fanning
Microbiology Department, University College Cork,
Cork, Ireland.

Summary:

A number of gene which can be used as convenient and reliable 'marker genes' in Rhizobium and Pseudomonas spp. has been studied. The availability of cloned lacZ and cya genes from R. meliloti in conjunction with transposon Tn951 and an 'integrating' genetamicin resistant vector provides the basis to allow the persistence and transfer of introduced genes to be followed in soil bacteria.
Key words: lacZ, Tn951, cya gene, integrating vector, R. meliloti, Pseudomonas spp.

Introduction

The evaluation of risk involved in the deliberate release of genetically engineered seed inoculum microorganisms into soil/rhizosphere environments is currently an important consideration in Biotechnology. The application of genetic engineering technology has major potential for the improvement of plant-microbial interactions, particularly in the

Risk Assessment for Deliberate Releases
Edited by W. Klingmüller
© Springer-Verlag Berlin Heidelberg 1988

context of improving traits for commercial application. In the Rhizobium-Legume symbiosis, traits of agronomic interest include host range and energy efficiency. Alterations of key genes controlling nodulation and carbon metabolism are the targets of research programmes in many laboratories around the globe (1, 2, 3, 4). For example, recent work on the manipulation of the dicarboxylic acid transport (dct) system in Bradyrhizobium japonicum using cloned dct genes has resulted in genetically engineered strains that show increased transport and 'in vitro' nitrogen fixation activity (4).

In another important area of plant-microbial associations, Pseudomonas spp. are being exploited for plant growth stimulation and biological control purposes (5, 6, 7, 8, 9). For the development of ideal inoculants in this category, strains with the following characteristics would be desirable:
(a) ability to inhibit a broad spectrum of fungal pathogens, (b) ability to agressively colonize root systems and (c) metabolic flexibility to compete and utilize root exudate materials as nutrients for their persistence in the rhizosphere. It is also evident that genetic engineering technology has the potential to assist in the creation of inoculant strains with these characteristics.

In the development and exploitation of biotechnology, concerns have been expressed regarding potentially harmful effects to mankind and the environment by some applications of genetic engineering in microorganisms. In relation to applications in industrial fermentations, genetically altered microorganisms are normally contained during research or manufacturing processes and can be destroyed when the particular fermentation run is completed. However, with genetically modified microorganisms intended for use in agriculture, the situation is more complex. The exchange of genetic material from the 'engineered organism' with other microorganisms is one of the main areas of concern. In attempting to evaluate the potential for gene transfer between soil microorganisms in pilot experiments, a prerequisite step is the availability of suitable 'marker genes' and strategies which can be exploited to monitor gene transfer unambiguously. In our programme, we are developing a number of genes which may be used as convenient and reliable monitoring systems in *Rhizobium meliloti* and *Pseudomonas* spp.

Materials and Methods

Bacterial strains and plasmids: *E. coli* MC1061, *ara*D139, Δ(ara-leu)7697, Δ(lacIPOZY)X74, *gal*U, *gal*K, *hsd*R, *hsd*M, *str*A, was obtained from M.J. Casadaban (10). *P. fluorescens* M11/4 was isolated in our laboratory (9). *R. meliloti* 102F34 was obtained from G. Ditta (11) and a spontaneous Rifampicin resistant mutant was isolated in our laboratory. Plasmid pGC91.14(RP1::Tn951) is described elsewhere (12), as is p2046 (13). Plasmid 246 (and its derivatives) were constructed in our laboratory by cloning an AAC(3)III gene from an *E. coli* clinical isolate into pBR322 (14).

Media and growth conditions: Complex medium for *E. coli* was Luria Bertani (LB) broth (15). *Pseudomonas* was maintained on sucrose asparagine medium (5) and *Rhizobium* species on mannitol salts yeast extract (MSY) (16). Minimal medium for *E. coli* was M9 (17). Levels of antibiotics used were: gentamicin 20 μg/ml; tetracycline 5 μg/ml (RP1::Tn951 in *Rhizobium*); kanamicin 25 μg/ml (pRK2013 in *E. coli*); Rifampicin 100 μg/ml (*R. meliloti* Rif^{r}).

Genetic techniques: Conjugation transfer of pBR322 based plasmids was done by a tri-parental mating with an *E. coli* strain bearing pRK2013 (11) as a helper. A spontaneous Rif resistant mutant of *R. meliloti* F34 was used as a recipient strain. Recombinant DNA techniques were done as described in (15).

Biochemical assays: β-galactosidase was assayed as described by Miller (18), using toluenized cells.

Results

Monitoring system for plasmid located genes

The design and testing of genetically engineered strains for improvements in specific traits may involve the introduction of desired gene(s) located on plasmids. In developing a model system to monitor gene transfer of this nature in *Rhizobium* and *Pseudomonas* species, we have exploited the lactose transposon Tn951 on plasmid RP1 (19). The main advantage of this system is that the *lac* genes provide a convenient reporter system to monitor transfer. Previous studies of the plasmid-borne Tn951 *lac* operon in *E. coli* have demonstrated a close similarity to the endogenote *E. coli lac* operon with respect to IPTG inducibility and sensitivity to glucose repression (20). The expression of Tn951 in *R. meliloti* was investigated under both basal and induced (IPTG) conditions. The induction ratio for β-galactosidase activity in *R. meliloti* has previously been reported to be 14.3 (19). In *R. meliloti* F34 lacking Tn951, β-galactosidase activity is not detected when IPTG is used as inducer. This observation provides the basis to rapidly identify *Rhizobium* clones carrying the transposon. This can be easily achieved in plate inductor tests employing IPTG and the β-galactosidase indicator 5-Bromo-4-chlor-3-indolyl-β-D-galactosidase (X-gal). The transposon can be monitored in either a plasmid or chromosomal location. In addition, the ability to monitor transmission to soil *Pseudomonas* spp. is also possible as fluorescent *Pseudomonas* strains are normally incapable of lactose utilization and give a negative response on X-gal indicator plates.

Novel *lac*Z gene for monitoring gene transfer

The desirability of using well characterized gene(s) such as the enteric *lac* operon as 'marker genes' to monitor gene transfer between microorganism is well recognised. However, a possible limitation associated with this system is the difficulty of distinguishing between transferred *lac* genes and resident *lac* genes in reservoir microorganism in soil environments. For example, DNA sequence conservation may make routine discrimination by DNA hybridization studies difficult. Recent work in our laboratory has resulted in identifying a *lac*Z gene from *R. meliloti* that displays unusual properties relative to the enteric *lac*Z gene and, therefore, provides the possibility to unambiguously monitor its transfer. The *R. meliloti* *lac*Z gene was isolated from a gene bank on a broad host range cosmid vector (pLAF.R1). By subcloning partial *Sau*IIIA fragment from the complementary *lac*Z cosmid clone into pBR322, a large number of sub-fragments were isolated. The smallest of these fragments had an insert of 2.4 kb, the coding region of the Rm *lac*Z gene (Fanning and

O'Gara, submitted). Using E. coli 'maxicells', a polypeptide of molecular weight 79 K-Da was identified. This polypeptide is significantly smaller than the corresponding protein activity found in E. coli. In DNA hybridization experiments, the cloned Rm lacZ gene did not show any sequence homology to the E. coli lacZ gene (unpublished data). This observation has potential applications in developing this sequence as a 'marker gene'. In particular, it provides the desirable trait of being easily monitored in bacteria by X-gal indicator plates and to be specifically identifiable from enteric lacZ type genes by DNA hybridization in colony blots.

Integration of genes into the chromosome

In the context of genetic modification applied to seed inoculants, the stability of introducted gene(s) is of fundamental important where commercial application is concerned. In practice, this situation dictates the development of strains where introduced genes are integrated into the chromosome and appropriate integrating vectors for the construction of such strains. In our programme we have exploited a cloned gene for adenyl cyclase (cya) (21) and a gentamicin resistance gene to construct an integration vector to target desirable genes into the genome of R. meliloti. A gentamicin resistance gene (AAC 3'III) from an R factor of clinical origin was cloned, characterized and exploited to construct a cloning vector based on the pBR322 replicon, (Boesten and O'Gara, in preparation). This vector (pCU246) features a multiple cloning site and the gentamicin resistance gene provides a 'clean' selectable marker for soil bacteria such as Rhizobium and Pseudomonas. This plasmid was further developed as an integration vector, exploiting the cloned Rm cya gene (21). Plasmid CU246 can be mobilized into R. meliloti but does not replicate in this organism, since it is based on a colEI origin of replication. However, plasmids containing a homologous region to the chromosome can become established through site specific integration. The site of integration becomes very important as disruption of a necessary gene could adversely affect the properties of the micro-organism. Recent work in our laboratory has identified a cya DNA sequence (21) that can be exploited to 'target' the integrating vector pCU246 to a site which does not result in cellular/metabolic disruption. The cloned sequence for adenyl cyclase (cya-1) from R. meliloti is apparently unique to R. meliloti (21) and the organism has an additional synthetic system for cAMP formation (Kiely and O'Gara, unpublished). Mutations or deletions in the cya-1 region do not result in a defective phenotype. Consequently, we have used this sequence cloned into pCU246 to test for site specific insertion (Table 1).

TABLE 1: Integration of vector pCU246 into the chromosome of R. meliloti based on cya chromosomal fragments (21) for homology.

Plasmid	cya Fragment	Size (kb)	Gm^r/Recipient Cell
p246	-	-	$< 8 \times 10^{-8}$
p246-1	EcoRI	5.2	3.2×10^{-7}
p246-2	BglII BamHI	2.6	1.9×10^{-7}
p246-3	PstI	0.67	2.1×10^{-8}

These results indicate that site specific integration can occur and is dependent on the size of the homologous fragment cloned into the pCU246 vector. The ensuing constructs showed similar growth characteristics on a variety of cultivation media, indicating that vector integration at this site does not disrupt cellular functions.

Discussion

In attempting to evaluate the potential for gene transfer between soil microorganisms, a prerequisite step is the availability of suitable marker genes to monitor gene transfer unambiguously. In our programme, we have developed a number of genes which can be used as convenient and reliable marker genes for use in Rhizobium and Pseudomonas spp. In evaluating the potential alteration(s) in the behaviour of microorganisms subjected to genetic manipulation the 'gene' of interest could be located on (a) transmissible and non-transmissible vectors and (b) transposable and non-transposable sequences integrated into the chromosome. Successful implementation of any strategy to monitor gene transfer is obviously dependent on the ability to easily identify donor, recipient and ensuing exconjugant strains formed as a result of gene transfer. Marker genes such as lacZ which provide options for positive selection are ideal candidates for this work. Transposon Tn951 encoding the entire lac operon has potential in this context. It provides some advantages compared to other transposons, such as Tn5 which depend on drug resistance markers for routine detection. Since many soil bacteria express rather high intrinsic antibiotic resistance levels, it could mask the detection of transposon encoded antibiotic resistance. The ability to selectively discriminate between the Tn951 encoded β-galactosidase and the endogenous activity in R. meliloti based on IPTG induction overcomes limitations of this nature when dealing with Tn951. However, as the expression of Tn951 is dependent on cAMP/CRP complex (19), the expression of β-galactosidase would need to be carefully monitored in a variety of Gram negative soil bacteria to ensure adequate levels of expression for widespread detection purposes.

The lacZ gene cloned from R. meliloti provides additional scope for exploiting this gene as a versatile marker gene. It has the advantage that it is smaller than the E. coli-type lacZ gene and it can be distinguished from it by lack of sequence homology. An additonal feature of this gene is that it can weakly complement E. coli mutants deleted in the entire lactose operon (Fanning and O'Gara, unpublished). It, therefore, may be possible to exploit this individual gene for a lactose positive phenotype without the necessity for the lacY (permease) gene. For example, current use of the E. coli lacZ as a marker gene in fluorescent Pseudomonads requires expression of the lacY gene product in these strains (22).

The development of strains where introduced genes are integrated into the chromosome is likely to increase in importance as agricultural biotechnology expands. In the case of R. meloliti the cya-1 sequence provides an ideal site for the stable integration of additional genetic information without disrupting cellular functions. The gentamicin resistance marker associated with the integrating vector provides a very useful positive selectable marker for soil bacteria. An alternative system for inserting a gene into the chromosome of soil bacteria has been reported and is based on a bicomponent transposition system between Tn7 termini (23). If the insertion of this system shows the same specificity as Tn7 in R. meliloti, then it would target insertions into a megaplasmid in these strains (24). Since the megaplasmids are involved in a number of symbiotic functions and can also be mobilized, it would not be a desirable site for permanent insertion of foreign genes. Consequently, the cya based integration vector would appear to offer a realistic system for integrating desired gene(s) in the R. meliloti genome.

Acknowledgements

We are grateful to the scientists listed in the materials and methods who provided us with strains and plasmids. The technical assistance of P. Higgins is greatly appreciated. This work was supported in part by grants from the EEC BAP Programme (Contract No. GB15-030 EIR), the NBST (SG1/84) and the Medical Research Council of Ireland.

References

1. Kondorosi, E. and Kondorosi, A. (1986). Trends in Biochem. Sci. 11, 269-299.
2. Halverson, L.J. and Stacey, G. (1986). Microbiol. Revs. 50, 193-225.
3. Ronson, C.W. and Astwood, P.M. (1985). In : Nitrogen fixation research progress, (ed.) H.J. Evans, P.J. Bottomley and W.E. Newton, pp. 201-207, Martinus Nijhoff Publishers, Dordrecht, Boston and Lancaster.
4. Birkenhead, K., Manian, S.S. and O'Gara, F. (1987). J. Bacteriol. (in press).
5. Scher, F.M. and Baker, R. (1982). Phytopathology 72, 1567-1573.
6. Weller, D.M. and Cook, R.J. (1983). Phytopathology 73, 463-469.
7. Schroth, M.N. and Hancock, J.G. (1983). Science 216, 1376-1382.
8. O'Gara, F., Treacy, P., O'Sullivan, D., O'Sullivan, M. and Higgins, P. (1986). In : Iron, siderphores and plant diseases, (ed.) T.R. Swinburne, pp. 331-339. Plenum Press, New York and London.
9. Stephens, P.M., O'Sullivan, M. and O'Gara, F. (1987). Appl. Environ. Microbiol. 53, 1164-1167.

10. Casadaban, M. and Cohen, S.N. (1980). J. Mol. Biol. 138, 179-207.

11. Ditta, G., Stanfield, S., Corbin, D. and Helinski, D.R. (1980). Proc. Natl. Acad. Sci. 77, 7347-7351.

12. Cornelis, G., Ghosal, D. and Seadler, H. (1978). Mol. Gen. Genet. 160, 215-224.

13. Kiely, B. and O'Gara, F. (1983). Mol. Gen. Genet. 192, 230-234.

14. Bolivar, F., Rodriguez, R.L., Greene, P.J., Betlach, M.C., Heyneker, H.C., Boyer, H.W., Crosa, J.H. and Falkow, S. (1977). Gene 2, 95-113.

15. Maniatis, T., Fritsch, E.F. and Sambrook, J. (1982). Molecular Cloning : A laboratory manual. Cold Spring Harbor Laboratory, Cold Spring Harbor, N.Y.

16. O'Gara, F., and Shanmugan, K.T. (1976). Biochem. Biophys. Acta 437, 313-321.

17. Figurski, D., Meyer, R., Miller, D.S. and Helinski, D.R. (1976). Gene 1, 107-119.

18. Miller, J.H. (1972). Experiments in molecular genetics. Cold Spring Harbor Laboratory, Cold Spring Harbor, N.Y.

19. McGetrick, A., O'Regan, M. and O'Gara, F. (1985). FEMS Microbiol. Letts. 29, 27-32.

20. Nano, F.E. and Kaplan, S. (1982). J. Bacteriol. 152, 924-927.

21. Lathigra, R., O'Regan, M., Kiely, B., Boesten, B. and O'Gara, F. (1986). Gene 44, 89-96.

22. Drahos, D.J., Hemming, B.C. and McPherson, S. (1986). Biotechnology 4, 439-444.

23. Barry, G.F. (1986). Biotechnology 4, 446-449.

24. Bolton, E., Glynn, P. and O'Gara, F. (1984). Mol. Gen. Genet. 193, 153-157.

SAFETY OF BACULOVIRUSES USED AS BIOLOGICAL INSECTICIDES

Jürg Huber
Biologische Bundesanstalt für Land- und Forstwirtschaft
Institut für biologische Schädlingsbekämpfung
Heinrichstr. 243, D-6100 Darmstadt, FRG

About 1100 viruses are known to infect insects, more than 60% of them being baculoviruses (1). The members of the virus family Baculoviridae are characterized by double stranded circular DNA, included in rod-shaped capsids which are formed mostly in the nucleus of the host cells. As a feature in common with some insect viruses from other virus families, the virions of most baculoviruses are contained within the matrix of proteinaceous particles, the so called occlusion bodies (OB), which provide protection against adverse physical and chemical factors in the environment. Based on the morphology of the OBs, baculoviruses are divided into 3 subgroups: (a) nuclear polyhedrosis viruses (NPV), where many virions are contained in an OB, (b) granulosis viruses (GV), where every virion has its own OB and (c) the small group of non-occluded baculoviruses, where no OB is formed (2). The natural way of infection is by ingestion of food contaminated with virus. In the gut of the susceptible host the matrix protein of the OB is dissolved and virus particles are relaesed. They enter the gut epithelial cells, where they multiply and subsequently quickly spread to several other tissues of the host. Baculoviruses do not produce toxins, but their massive multiplication in vital tissues eventually leads to the death of the insect, usually within 1 to 2 weeks after infection.

Host range of baculoviruses

The rapidly expanding discipline of Invertebrate Pathology has failed till today, to find baculoviruses in organisms other than members of the Arthropoda.

Risk Assessment for Deliberate Releases
Edited by W. Klingmüller
© Springer-Verlag Berlin Heidelberg 1988

This lack of evidence for the occurence of baculoviruses in non-arthropod host is sustained by the absence of baculovirus incidence from the publications of medical, veterinary, and phytopathology science. Baculoviruses have been isolated from crustacea and mites, but mostly from insects. The individual members of this virus family have a very narrow host range. They infect only a few species closely related to - and seldom outside the family of - the host species in which they were first found. The baculovirus with the widest host range known today is the NPV of Autographa californica (AcNPV). It is reported to infect larvae from about a dozen lepidopteran families - at least in the laboratory (3). It's pathogenicity in the new host usually is much lower than that of the corresponding homologous virus for a given host. Therefore, in spite of its wide host range, the AcNPV has never been commercialized as a viral pesticide. The other extreme with regard to host range are baculoviurses as, e.g., many NPVs from sawflies, for which no transmission to another host, not even from the same genus, is possible. The high selectivity makes baculoviruses ideal tools for integrated pest control strategies. Their narrow host range rates them amongst the environmentally least disruptive pesticides available at present (4).

Baculoviruses and environment

Baculoviruses are natural components of many ecosystems. Quite often, they are a controlling factor in the mass outbreak of insect populations. There are several reports indicating that after application of virus for control of a pest population, the virus load in the environment is in the same size or even smaller than after the collapse of the population through a late infection by indigenous virus (5,6).

The environmental stability of baculoviruses has been well studied (for review see 7). It is known that by far the most important inactivating factor is the UV radiation of the sun. Half-lives of one to two days appear common for foliar deposits of unprotected virus. In the ground, baculouviruses are retained within the first few centimeters of the top levels of the soil. Since they are protected from the sunlight, they can persist there for several years. Soil borne virus seems to be the main source for the initiation of viral epizootics in nature.

Viral insecticides

Worldwide, about a dozen baculoviruses are registered for use as biological insecticides in forestry and agriculture (8). This number seems small with regard to their considerable potential. It has been estimated that "baculoviruses can be used against about 30% of the major pest species interfering with food and fiber production in the world" (9). One of the major obstacles for a broader commercialization of viral insecticides is the economic descrepancy between the small potential market for such a selective product on one hand, and the high financial effort that has to be made for its registration on the other hand.

In spite of the indirect evidence for the safety of baculoviruses for man and environment by their absence from non-arthropod organisms, they had to undergo a series of tests, sometimes more demanding than those required for chemicals, because they were examined not only for toxicity but also for the possibility of infectivity for non-target organisms. These tests included long term carcinogenicity and teratogenicity studies on mammals and even primates and men. In all assays baculoviruses proved harmless to - and unable to replicate in - microorganisms, non-insect invertebrate and vertebrate cells, plants, non-arthropod invertebrates and vertebrates (10).

Only recently, genetic engineering has brought forward a new aspect of baculoviruses, which activated the interest of private industry in them. OB-forming baculoviruses have a number of unique features which make them very attractive vectors for the expression of foreign genes in an eukaryotic environment. The feasibility of this expression vector system has been proved by the production of fully functional human interferon, interleucin and other polypeptides in vitro in insect cells and even in vivo in live insects, using the strong promotor of the OB-protein of the AcNPV (for review see 12).

Genetically engineered baculoviruses

The research on molecular biology of baculoviruses, stimulated by these novel and fascinating prospects, lead to considerations to use genetic engineering technology also in order to improve their pesticidal qualities. Applying this technology, it should be possible to produce viruses which (a) exhibit a broader host range, (b) produce a toxin for faster kill, (c) show enhanced virulence, or (d) have a better environmental stability than their naturally occuring "parent" viruses. Since these genetically manipulated viruses are not going to be contained in the laboratory as in the case of expression vector systems, but are meant for release into the environment, their safety has to be fully tested. Hardly any experiecences have been collected up to now in this novel field of risk assessment. In many aspects the dissemination of such an "artificial" virus resembles the introduction of a naturally occuring virus into a new environment from which it has been absent so far (e.g., releasing an "exotic" virus isolated in Australia in a European orchard).

Most of the goals of genetic engineering technology listed above could also be envisaged by conventional methods. It has, e.g., been attempted several times to adapt a virus to a new host by serial passages in the other insect in order to have a substistute host for the mass production of the virus or to have a new viral insecticide against a given pest (13). Similar successfully, it has been tried to increase the virulence of a virus against its host by repeated passages in it (14) or even improve its resistance against UV-radiation (15). Obviously, the outcome of all these trials was mostly governed by chance since it depended on the genetic situations at the beginning of the selection process. In some trials, attempts were made to increase this genetic basis by the addition of chemical mutagens prior to, or during the infection of the larvae (16, 17). With regard to safety, these viruses were treated the same as the naturally occuring "parent" viruses.

Unfortunately, hardly anything is known about the biological and biochemical mechanisms reponsible for the high specificity of baculoviruses. Recent studies have shown that the block for multiplication of baculoviruses in non-permissive cells occurs not at the level of virion binding to the plasma membrane but in the nucleus, probably because of inproper uncoating of the virion or the failure of the cells to transcribe the viral DNA (18). Nothing is known about the genetic basis for the specificity of baculoviruses. Therefore, in the near future, also genetic engineering technology would have to rely mostly on chance when trying to alter the host range or the virulence of baculoviruses.

Incorporation of new genes, responsible for the production of toxins into the virus genome as, e.g., the gene for the δ-endotoxin of Bacillus thuringiensis seems more feasible. The gene encoding this toxin is well known and has already been inserted into tomato and tobacco plants where its expression rendered the plants toxic for insects feeding on them. The gene, integrated into baculovirus genome, could improve the rather slow action of the natural virus. Its toxin could induce a stop of the feeding activity of the insect at an early stage of infection. At the same time it would reduce the possible hazard of the virus to multiply uncontrolled in the environment. The early death induced by the toxin would prevent the maturation of the viruses.

Risk assessment

At present, there seems to be no need for totally new safety testing schemes for genetically engineered viruses. The difference between "natural" and modified viruses is rather a quantitative than a qualitative one. The potential hazards forseen with genetically engineered viruses are, for most part, the same as for non-engineered microbes (19). A major concern is the potential risk of transfer and exchange of genetically engineered characteristics with naturally occuring microorganisms, or with the host or non-target organisms, thereby creating new potential hazards. It is known that baculoviruses do form recombinants in vivo with other closely related baculoviruses (20), but even the baculovirus with the widest host range, the AcNPV, was not found to persist in non-permissive (e.g., mammalian) cells (21, 18).

For the near future, until some experience has been gained, each projected release of genetically engineered baculoviruses should be examined carefully.

Data from other baculoviruses cannot necessarily - as it had been sound practice with naturally occuring viruses in the past years - be taken as evicence for the safety of new viruses. The requirement for additional safety data should be determined on a case-by-case basis, depending on the particular virus, its parent viruses, and the manner and extent to which the virus has been genetically modified.

References

1. Martignoni, E.M. and Iwai, P.J. (1986): A catalog of viral diseases of insects, mites and ticks. Gen. Tech. Rep. PNW-195, USDA, Forest Service, 51 p.
2. Matthews, R.E.F. (1982): Classification and nomenclature of viruses. Intervirol. 12, 1-199.
3. Payne, C.C. (1986): Insect pathogenic viruses as pest control agents. Fortschr. Zoolog. 32, 183-200.
4. Podgewaite, J.D. (1986): Effects of insect pathogens on the environment. Fortschr. Zoolog. 32, 279-287.
5. Jaques, R.P. (1974): Occurrence and accumulation of the granulosis virus of Pieris rapae in treated field plots. J. Invertebr. Pathol. 23, 351-359.
6. Podgewaite, J.D., Stone Shields, K., Zerillo, R.T. and Bruen, R.B. (1979): Environmental persistence of the nucleopolyhedrosis virus of the gypsy moth Lymantria dispar. Environ. Entomol. 8, 528-536.
7. Jaques, R.P. (1977): Stability of entomophathogenic viruses. Misc. Publ. Entomol. Soc. Am. 10(3), 99-116.
8. Huber, J. (1986): Use of baculoviruses in pest management programs. In: The Biology of Baculoviruses. Vol. II. Practical Application for Insect Control. (Eds. Granados, R.R. and Federici, B.A.), pp. 181-202, CRC Press, Boca Raton.
9. Falcon, L.A. (1980): Economical and biological importance of baculoviruses as alternatives to chemical pesticides. Proc. Symp. "Safety Aspects of Baculoviruses as Biological Insecticides". 13.-15. Nov. 1978, Jülich, 27-46.
10. Burges, H.D., Croizier, G. and Huber, J. (1980): A review of safety tests on baculoviruses. Entomophaga 25, 329-340.

11. Gröner, A. (1986): Specificity and safety of baculoviruses. In: The Biology of Baculoviruses. Vol. I. Biological Properties and Molecular Biology (Eds. Granados, R.R. and Federici, B.A.), pp. 177-202, CRS Press, Boca Raton.

12. Doerfler, W. (1986): Expression of the Autographa californica nuclear polyhedrosis virus genome in insect cells: Homologous viral and heterologous vertebrate genes - The baculovirus vector system. Curr. Top. Microbiol. Immunol. 131, 51-68.

13. Martingoni, M.E. and Iwai, P.I. (1986): Propagation of multicapsid nuclear polyhedrosis virus of Orgyia pseudotsugata in larvae of Trichoplusia ni. J. Invertebr. Pathol. 47, 32-41.

14. Veber, I. (1964): Virulence of an insect virus increased by repeated passages. Entomophaga, Mém. hors. Sér. 2, 403-405.

15. Brassel, J. and Benz, G. (1979): Selection of a strain of the granulosis virus of the codling moth with improved resistence against artificial ultraviolet radiation and sunlight. J. Invertebr. Pathol. 33, 358-363.

16. Wood, H.A., Hughes, P.R., Johnston, L.B. and Longridge, W.H.R. (1981): Increased virulence of Autographa californica nuclear polyhedrosis virus by mutagenesis. J. Invertebr. Pathol. 38, 236-241.

17. Mc Clintock, J.T. and Reichelderfer, C.F. (1985): In vivo treatment of a nuclear polyhedrosis virus of Autographa californica (Lepidoptera: Noctuidae) with chemical mutagens: Determination of changes in virulence in four Lepidopteran hosts. Environ. Entomol. 14, 691-695.

18. Brusca, J., Summers, M., Couch, J. and Courtney, L. (1986): Autographa californica nuclear polyhedrosis virus efficiently enters but does not replicate in poikilothermic vertebrate cells. Intervirol. 26, 207-222.

19. Betz, F.S. (1986): Registration of baculoviruses as pesticides. In: The Biology of Baculoviruses. Vol. II. Practical Application for Insect Control (Eds. Granados, R.R. and Federici, B.A.) pp. 203-222, CRC Press, Boca Raton.

20. Croizier, G. and Quiot, J.M. (1981): Obtention and analysis of two genetic recombinants of baculoviruses of Lepidoptera, Autographa californica and Galleria mellonella. Ann. Virol. 132, 3-18.

21. Tjia, S.T., Meyer zu Altenschildesche, G. and Doerfler, W. (1983): Autographa californica nuclear polyhedrosis virus (AcNPV) DNA does not persist in mass cultures of mammalian cells. Virol. 125, 107-117.

AN OVERVIEW OF INSECT BACULOVIRUS ECOLOGY AS A BACKGROUND TO FIELD RELEASE OF A GENETICALLY MANIPULATED NUCLEAR POLYHEDROSIS VIRUS

P.F. Entwistle, J.S. Cory and C. Doyle
Natural Environment Research Council, Institute of Virology
Mansfield Road, Oxford OX1 3SR, UK

Summary: The use of baculoviruses (BVs) as bio-insecticides could be enhanced by incorporation of genes leading to reduced damage to crops by the infected host insect during the period of disease incubation. The risks associated with release of genetically engineered BVs can be best assessed against a comprehensive background of knowledge of their natural ecology. This paper summarises what is known of BV ecology and presents a structured programme of further research employing a virus belonging to this group which has been genomically labelled with a non-coding synthetic oligonucleotide sequence. Such a marker would permit an unique BV population to be traced through the environment in space and time.

The role of ecological studies in assessing risks associated with genetically engineered baculoviruses.

Genetically unmodified Baculoviridae (BVs) have a demonstrable potential as insect bio-control agents, not only because of their innate insecticidal potency but also because they present few environmental problems. Their use is especially promising against larval Lepidoptera and diprionid sawflies (Hymenoptera). However, there are some disadvantages, as pest feeding during the BV-disease incubation period can result in unacceptable crop damage. Furthermore, the mortality response declines strongly with host larval age making careful timing of BV application desirable for economic reasons.

It is postulated such disadvantages could be obviated by the incorporation into BVs of appropriate genes. These could be 'killer' genes inducing swifter mortality (e.g., the D-endotoxin gene of *Bacillus thuringiensis*), genes which strongly inhibit feeding or which advantageously modify pest behaviour.

Risk Assessment for Deliberate Releases
Edited by W. Klingmüller
© Springer-Verlag Berlin Heidelberg 1988

Before environmental release of a genetically modified BV an assessment must be made of the risks involved. The current world view is that such risks can be judged only against a background of much improved understanding of their ecology. This attitude was recently endorsed in a meeting considering the Basic Research Needs in Microbial Ecology in an Era of Genetic Engineering (1) and was further expressed in the 1987 Biotechnology Environmental Release Database Workshop (2): a strong need was recognised to include information on both the biology and ecology of wild types as well as those genetically modified.

Our project is in accord with such opinions. We seek initially to address basic questions of BV ecology employing a BV labelled by insertion of a non-coding synthetic sequence of oligonucleotides (the marker 'gene') but which is otherwise unaltered. In combination with marker 'gene' - specific DNA probes, this will permit the environmental fate of the virus to be followed. The ecological behaviour of such a virus will, as far as possible, be compared with the wild type from which it was derived.

The nature of baculoviruses

More than ten families of viruses have been identifed causing infections in insects. The virus family Baculoviridae is uniquely associated with the Insecta and some other Arthropoda, notably paenid shrimps. The rod-shaped virions (*circa* 200 nm in length) contain a circular double stranded DNA molecule. Three subgroups are recognised: A, the nuclear polyhedrosis viruses (NPVs); B, the granulosis viruses (GVs) and, C, the *Oryctes* type viruses. In subgroups A and B the genome codes for a protein which forms occlusion bodies around the virions, incorporating many virions in A and a single virion in B. The NPV occlusion body is polyhedral in shape (0.8 - > 5 µm diameter) but in GVs is generally elipsoidal (0.3 - 0.5 µm long). Subgroup C viruses are not occluded. NPVs are largely found in Lepidoptera but some affect other taxa, notably Hymenoptera-Symphyta. GVs are exclusively associated with Lepidoptera but subgroup C viruses are found in a wide range of host taxa. Positive infections by NPVs and GVs are largely restricted to host larvae and are mainly initiated by virus ingestion. In so far as is known, subgroup C viruses can be infectious in larval and adult host stages (3).

i. BV environmental significance

BVs are natural regulators of insect populations and catastrophic epizootics are commonplace both in nature and in cultivated field and forest crops. When infected hosts die and decay much virus is liberated into the environment. Such virus is the principal source of infection and in subgroups A and B has considerable environmental stability bestowed by the occlusion body. Subgroup C viruses are less stable and their transmission depends on more immediate association of diseased or dead individuals with healthy hosts. The coevolution of virus and host has resulted in various patterns of virus strategy. These have been explored in relation to examples of hosts occupying different regions in the r-K continuum (4, 3).

ii. BVs as pest control agents.

The use of BVs as pest control agents relates especially to subgroups A and B where the occlusion bodies provide stability of virions in storage making them valuable sprayable bio-insecticides. Several BVs have been developed to a commercial level and many others have been shown to have such potential (5). BVs have been shown to be harmless to man, other vertebrates and plants (Huber, this publication). Beneficial insects (parasites and predators) are unaffected and in this and other respects BVs are very much less environmentally injurious than are chemical insecticides. The majority of BVs have a narrow host spectrum and so pose little threat to insects of conservation interest or to those which are major components of animal food chains and webs (5).

Contemporary BV ecology

Epizootics are cyclical events in which, with distinctive rates and patterns, infection changes quantitatively in space and time - components which, for purposes of discussion, can be considered separately. The speed of completion of an epizootic depends on the continuity of host generations, normally faster in tropical than in temperate regions. An epizootic usually occupies several host generations to full expression.

i. Temporal growth of disease

Infection follows a wave form in lag phase with changes in host population: this is the well known temporal infection wave. It encompasses several host generations in each of which the rate of infection is a reflection of host and virus population densities. Fluctuations in levels of these have a strong governing role on temporal epizootic rates.

ii. Spatial growth of disease

The spread of BV disease from an initially small epicentre appears to follow uniform patterns, the rate-expression of which can be individually characteristic for different virus-host systems. Disease spread appears typically to follow three identifiable phases. In the primary phase the relationship of the distance of disease spread (x) to the proportion of the population diseased (y) follows an indented curve. Normalisation by log conversion of the units of both axes provides a characteristic rate (-b) of spread. Subsequently, following high epicentral mortality, disease dispersal evolves to a wave form, the spatial dispersal wave phase. At this early stage of epicentral infection growth, it is apparent any cross-section of the spatial disease developmental process tends to express the whole or part of the temporal infection wave. Thus, epizootic growth in time and space are inseparable processes. In the third, or confusion phase, the clarity of the early pattern of epicentral growth tends to be lost through interaction with secondarily developing disease centres (6).

iii. Mediation of tempero-spatial epizootic patterns

The development of tempero-spatial pattens of epizootic change, though modified by host infection responses, are entirely controlled by the effectiveness of persistence and dispersal mechanisms.

The principal areas of NPV and GV persistence are the soil and the plant surface. (Persistence in sympatric alternative host species has not yet been investigated.) Persistence in soil can be measured in terms of years. The rate of return of BVs from soil to susceptible insects is probably low so that soils may be more important as a long term virus bank deposit account than as an intermediate important influence on the epizootic course. BVs can persist physically on plant surfaces for up to two or possibly more years. However, on the more exposed surfaces they are rapidly inactivated by solar UV B radiation. Not surprisingly, biological persistence appears to increase with increasing complexity of plant architecture so that it is less on annual field crops and much more on trees. Quantities of NPV persisting overwinter on coniferous trees may govern epizootic dynamics (7).

Spatial growth of epizootics depends on the levels of effective abiotic and biotic dispersal of inoculum. As far as is known abiotic dispersal (wind and rain) has only a local influence, except possibly in drier areas where BV occlusion bodies could be appreciably wind-borne in dust. Biotic dispersal is better documented, at least in a qualitative sense. 'Vectors' can be mobile life stages of the host itself (e.g., adult diprionids with active NPV gut infections, pre-infected wind-borne early instar larvae of some flightless Lepidoptera such as *Lymantria dispar*). Predatory and parasitic animals may be more generally important. NPVs and GVs are known to pass unaffected through the alimentary canals of almost all insectivorous animals whether they be invertebrates (e.g., Coccinelidae, Carabidae, Neuroptera, Reduviidae [Insecta], Opilionidae [Arachnoidea]) or vertebrates (Soricidae, Cricetidae, Muridae and Aves). Earthworms (Lumbricidae) which also have this capacity, are saprophagous and may ingest BVs on fallen plant parts.

Because of their strong searching ability, voraciousness and mobility, birds may be particularly important vectors. A study of a *Gilpinia hercyniae* NPV disease epizootic in Welsh spruce forests in 1974 showed 16 bird species and over 90 percent of individuals carried the virus, whilst over 70 percent of birds caught in Scotland in late July 1986 in a 320 hectare pine forest sprayed early in the preceding month carried NPV presumed to have been acquired from infected larvae of *Panolis flammea*.

Parasitic Hymenoptera, and probably to a lesser extent Diptera, may also vector BVs and have been shown to be able to infect several insects sequentially following contact with a single diseased individual. The carriage of BVs in forest systems has been reviewed (8).

The components of the selected study systems

The virus selected for study is an NPV originally isolated from the Cabbage moth, *Mamestra brassicae* (MbNPV) but cross infective to the related *Panolis flammea*, Pine beauty moth and to a range of other Lepidoptera. A synthetic non-coding sequence of oligonucleotides is being inserted into clones (designated MbNPV m) selected for ecological study. The ecology of wild type MbNPV has already received some attention (9, 10) in *M.brassicae* in a cabbage crop field ecosystem and the virus has been extensively sprayed in *P.flammea* field control trials in Scotland where studies on the ecology of the closely homologous PfNPV are in progress.

Studies on the ecology of MbNPV fall into two phases. Firstly, various questions will be addressed employing simulated ecosystems in severe containment facilities in the laboratory. This work will involve a strong ecological contrast between *M.brassicae* in a cabbage crop system (Institut für biologische Schadlingsbekampfung, Fed. Republ. Germany) and *P.flammea* in a pine forest system (Institute of Virology, UK). In the second phase, studies will extend to natural pine forest in the UK where it is intended to make a single release of MbNPVm into a *P.flammea* population. The results of current ecological studies on wild type PfNPV will indicate the optimal methods of environmental scrutiny to be employed with this release.

Assessing BV environmental pathways

It is not easy to escape the conclusion that the possible environmental pathways, especially for viruses with strong qualities of persistence such as NPVs, can be complex. For purposes of defining key questions for interrogation within contained ecosystems it is first necessary to analyse the possible patterns of environmental flow of NPVs. Such a flow picture is outlined in Figure 1. Three environmental regions are recognised; plant, soil surface and soil. From virus reservoirs accumulated within this putative system, abiotic and biotic vectors mediate virus flow within and between these regions.

The abiotic mechanisms of NPV dispersal are probably self-evident. To identify and classify real biotic pathways, however, it is necessary to define which animal taxa are associated with virus dispersal within each region and which between regions. In Figure 2 six categories of activity are identified and animal groups apportioned to them. Which of these categories are significantly represented in any 'real' ecosystem (cabbage fields, pine forests) can be determined only by field survey. In identifying which 'vectors' warrant study in containment facilities a balance of judgement should be achieved between pure numbers of a species, its biomass and its activity level. Thus, in pine forests bird populations are low, but relative biomass and activity are high compared, for instance, with opilionids which have higher populations but lower biomass and, being poikilothermic, are less active.

Interactions with alternative host species will be studied both in terms of their relevance in the flow of whole MbNPV genomes and also with a view to possible recombination events between this and heterologous NPVs.

Questions probed in a contained, simulated, pine forest ecosystem will provide guidance to further investigation in the natural forest ecosystem. Given sanction by the relevant authorities, these will be closely investigated following a single, unique, release of MbNPVm in a Scottish forest. This virus will be liberated by an introduction of *P.flammea* larvae previously infected in the laboratory into a natural population. It is of critical importance to an effective study that the size of such a release be adequate to ensure development of an NPV inoculum body large enough to withstand overwinter attritional forces and to have an infection impact on the following host generation. Therefore the relationship of the quantity of inoculum per unit area of forest produced in one host generation to levels of infection in the succeeding generation is currently being studied.

References

1. Regal, P.J., Klug, M., Sayler, G., Shapiro, J. and Tiedje, J. (1987). Nat. Sci. Foundation Executive Rpt. of April-May 1986, Scottsdale, Arizona Workshop. 18pp.
2. Anon (1987). Proceedings of a biotechnology environmental release database workshop. 95pp. Lister Hill National Center for Biomedical Communications, National Library of Medicine, National Institute of Health, Washington.
3. Evans, H.F. and Entwistle, P.F. (1987). In: Epizootiology of insect diseases (Eds. Fuxa, J.R. and Tanada, Y.) pp. 257-322, John Wiley and Sons.
4. Entwistle, P.F. (1986). In: Biological Plant and Health Protection (Eds. Franz, J.M. and Lindaur, M.) pp. 257-278, Fischer Verlag, Stuttgart.
5. Entwistle, P.F. and Evans, H.F. (1985). In: Comprehensive insect physiology, biochemistry and pharmacology (Eds. Kerkut, G.A. and Gilbert, L.I.) pp. 347-412, Pergamon Press, Oxford.
6. Entwistle, P.F., Adams, P.H.W., Evans, H.F. and Rivers, C.F. (1983). J. appl. Ecol., 20, 473-489.
7. Evans, H.F. and Entwistle, P.F. (1982). In: Microbial and Viral Pesticides (Ed. Kurstak, E.) pp. 449-461, Marcel Dekker, New York.
8. Entwistle, P.F. (1982). Proc. IIIrd Internat. Colloqu. Invertebr. Pathol., pp. 344-351, Brighton, UK.
9. Evans, H.F. (1982). Proc. IIIrd Internat. Colloqu. Invertebr. Pathol., pp. 307-312, Brighton, UK.
10. Evans, H.F. and Allaway, G.P. (1983). Appl. Environment Microbiol., 45, 493-501.

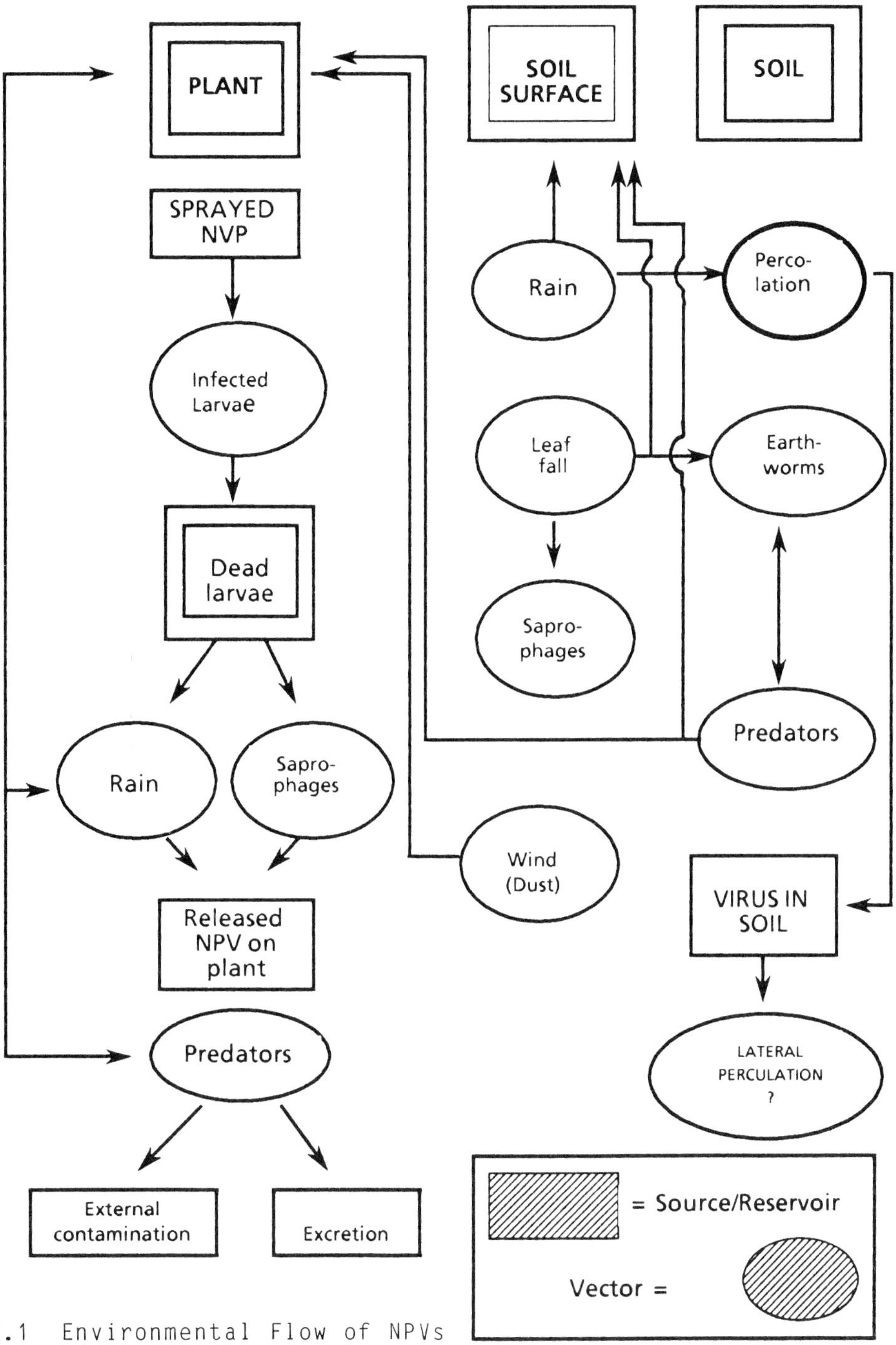

Fig.1 Environmental Flow of NPVs

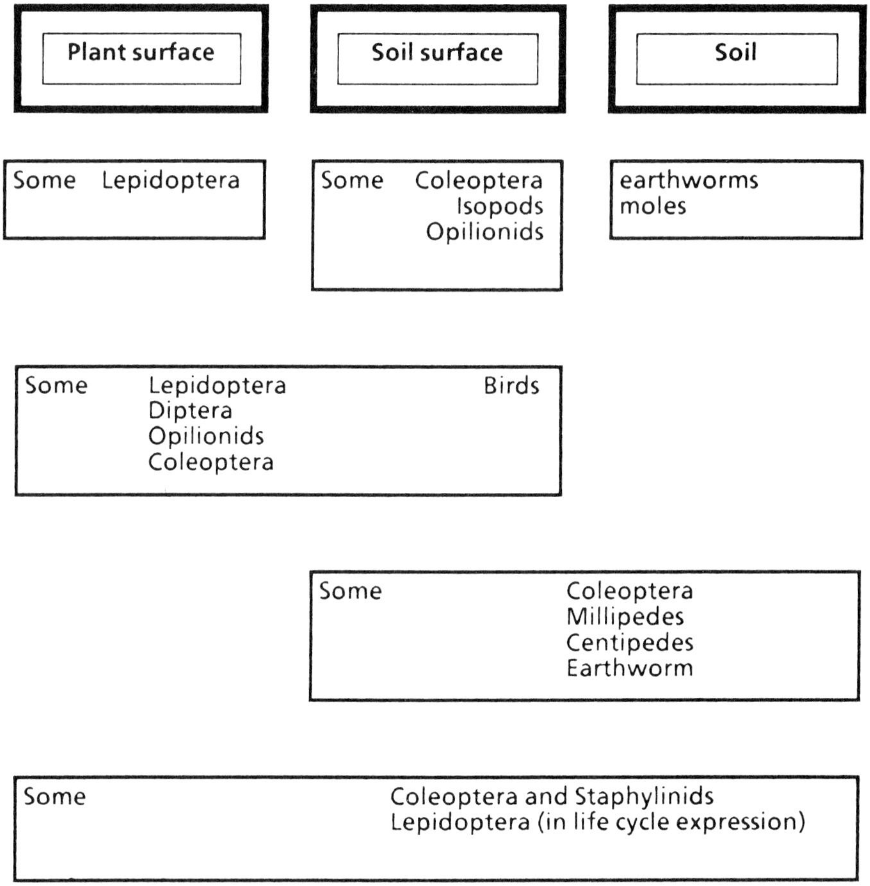

Fig.2 An analysis of NPV biotic vectors

SOME ECOLOGICAL ASPECTS OF THE RELEASE OF NONRESIDENT MICROORGANISMS IN SOIL AND GROUNDWATER ENVIRONMENTS

Z. Filip
Institut für Wasser-, Boden- und Lufthygiene
des Bundesgesundheitsamtes, Aussenstelle Langen
Heinrich-Hertz-Str. 29, D-6070 Langen, FRG

Summary: Nonresident microorganisms including genetically engineered ones are exposed to different biotic and abiotic factors when released into natural environments. In soils and groundwater aquifers, particulate materials may influence their survival, multiplication, and distribution. Indirectly, these materials can also influence gene transfer by conjugation, transformation and transduction. Some examples are discussed, and needs for further research are indicated in this paper.

Introduction

During a panel discussion on vulnerability of ecosystems to engineered microorganisms at the Conference on Genetic Control of Environmental Pollutants in 1983, in Seattle, Washington, a scientist said, "I have trouble understanding what happens in my petri dishes in the laboratory where I think I can control many things. What is happening in estuaries and soils is an enormous enigma; it is very, very complicated" (1). Some years ago I wrote that the time comes to transfer the copious information obtained in general microbiology into conditions of natural environments, and to further enlarge them in this way (2). Yet, though microorganisms in soil and water have been generally recognized as accomplishing manifold ecologically important functions in the total biosphere (e.g., nutrient cycling), genetic engineering has been the force which has turned serious attention towards microbial ecology. This is because

Risk Assessment for Deliberate Releases
Edited by W. Klingmüller
© Springer-Verlag Berlin Heidelberg 1988

a major concern is whether the release of genetically engineered microorganisms, to control pests and chemical pollutants, can cause harm to the existing homeostasis of terrestrial and aquatic environments. To study the fate of nonresident and especially genetically engineered microorganisms in soil and groundwater, the main microecological factors are organic and inorganic constituents. In natural environments they must be considered both for their existing importance in microbial ecology, and their possible role to the genetic material.

Natural distribution of microorganisms in soil and its possible consequences in genetic exchange

Soil is a matrix of colloidal systems. A full understanding of the microenvironments requires an examination of the characteristics of both colloidal particles and the matrix in relation to microbial life (3). In a soil sample, organic and inorganic particles, soil moisture, and gaseous phases, together with microorganisms, form a natural complex with numerous extensive interactions between the individual components. In this complex, clay minerals, and especially smectites and micas, strongly influence the physico-chemical properties of soil. They also effect microorganisms via adhesion of individual cells, and adsorption of nutrients, enzymes and different metabolic products (4, 5, 6).

When examining a soil suspension by either light or electron microscopy, mineral particles differ in their size and shape, and appear to be surrounded by a monolayer or by multilayers of bacterial cells, which are mainly cocci and short rods (7, and unpublished). Colloidal soil minerals, i.e., less than 0.2 μm in diameter, can be observed to adhere on bacterial and fungal cells. Several cultivation and biochemical methods, including ATP estimations, also demonstrated very close relationships between the soil organic and mineral particles, their size, and their inhabitance by different groups of microorganisms and microbial biomass (8). Since the physical contact between donor and acceptor cells is a necessity if genes are to move among microbial populations by conjugal mechanisms, it is clear that solid soil particles may influence this process by supporting the accumulation of bacterial cells on their surfaces. Such an effect has already been observed in soil samples enriched in montmorillonite, which resulted in an enhanced frequency of conjugal transfer of chromosomal genes from prototrophic strains of Escherichia coli to auxotrophic ones (9). However, the transport was significantly higher in sterile than in nonsterile soils, indicating that the presence of indigenous soil microorganisms interfered with this process.

Transformation is another mechanism of gene transfer which can be supported by the solid soil constituents. Several authors have demonstrated the adsorption of DNA and RNA to soil particulates, and a higher resistance of the adsorbed substances to enzymatic degradation (see in 4, and 10). When DNA was added to an experimental system containing quartz sand, the frequency of transformation of Bacillus subtilis was enhanced, presumably because the DNA was protected by adsorption (11).

A question arises, as to whether the adsorption and adhesion phenomena may be applicable in the natural environments where active surfaces and interfaces are already occupied by indigenous microorganisms or molecules. The answer is yes. Due to soil dynamics, ion layers which control the interfacial forces both on cell walls and abiotic surfaces may change considerably under natural soil conditions, resulting in a reciprocal adsorption and desorption of molecules, and adhesion and liberation of microbial cells. In model experiments, additional adhesion of Bacillus mycoides was observed on particle surfaces colonized in part by Serratia marcescens, while an exchangeable adhesion of both species occurred on surfaces fully colonized by S. marcescens (12). In both cases, the bacterium newly introduced into the experimental system came in a close contact with the 'indigenous' species, and there is no doubt that this phenomenon may also have consequences if genetically engineered organisms are involved. In different cultures of soil fungi adsorption and desorption of organic molecules by a clay mineral (montmorillonite) was again demonstrated using X-ray diffractometry (2).

Growth of microorganisms as influenced by particulate materials, and some possible implications to gene transfer

The growth activity, utilization of substrate, and yield of biomass in microbial cultures are greatly influenced by the presence of different particulate materials such as clay minerals, other silicates and humic substances naturally occurred in the terrestrial and aquatic environments. Positive effects have been reported for complex populations of soil microorganisms, and also for different pure cultures, especially for slowly growing fungal strains of *Aspergillus niger* (13). Colony counts and biomass yields increased up to ten times in comparison to the respective controls. A strong stimulation has also been observed with other than soil microorganisms, e.g., with culture collection strains of *Candida utilis* and *Saccharomyces cerevisiae*. This indicates that under proper conditions the growth of organisms other than those already adapted to the specific soil ecological factors may become stimulated. An enhanced growth activity, however, results also in an increased probability of gene transfer by conjugation simply because of larger numbers of donor and recipients cells. Furthermore, an enhanced release of DNA from the growing cells can be expected to result in an enhanced possibility of gene transfer by transformation. Further transformation can occur when an appreciable amount of DNA becomes liberated (and protected by solids) after cell lysis.

Survival of microorganisms as affected by particulate materials, and some possible implications to gene transfer

The survival of microorganisms in the natural environments depends not only on biological factors specific to the individual species but also on numerous biotic and abiotic ecological factors. In a recent review, general survival strategies of bacteria have been discussed (14). For genetically manipulated microorganisms, the survival is the main condition which controls possible gene transfer in the natural environments. Since few studies exist which deal with the survival of genetically engineered bacteria such as Pseudomonas spp., and E. coli (15, 16), it is necessary to consider the survival of different selected microorganisms nonresident to the soil and aquatic envinronments as a model for the fate of genetically engineered ones.

In our studies, some human pathogenic and facultatively pathogenic microorganisms and enteroviruses were tested for their survival in groundwater at 10° C. Most were found to decline in counts by up to several orders of magnitude during 300 day experiments. However, appreciable concentrations remain viable and reproductive both in the presence and absence of a mineral substratum (quartz sand) originating from a groundwater aquifer (17). This mineral substratum adsorbed both gram-negative and gram-positive bacteria, and viruses, and prevented their transportation in sand columns. This behaviour, i.e., a long survival and extensive spatial concentration on the surfaces of particulate materials will undoubtedly support gene transfer among the participatimg microorganisms. The long persistence of viruses and their remaining viability indicate that transduction as a gene transport mechanism also must be taken into account under natural conditions.

Both mineral and organic particulates have also been shown to protect microorganisms against a heat stress, and perhaps to support a population shift within the complex microflora (18). These effects can be of importance for survival of novel microorganisms. Similarly, the survival of engineered genes in the environment may be increased by the transfer to indigenous hosts that may be more fit for survival in a specific habitat (19).

Conclusions and needs for further research

From what has been said, one can conclude that genetically engineered microorganisms once released into the natural environment theoretically may have a good chance to survive, multiply spread,and cause gene transfer to an extent, as yet, not well known. In terrestrial and aquatic environments, abiotic factors such as inorganic and organic particulates may exert several supporting effects as mentioned above. To determine limits for these processes an urgent need exists for research in this field. With respect to the behaviour of genetically engineered microorganisms upon the soil and groundwater conditions, the following tasks need to be undertaken:
(i) Development of methods for identification and quantification of novel microorganisms in these complex environments;
(ii) Determination of survival, multiplication activity and transportation;
(iii) Determination of primary ways of gene transfer, i.e., conjugation, transformation, and transduction, with respect to the specific 'in situ' conditions;
(iv) Establishment of potential effects of recombinants on the diversity of indigenous species, their physiological activity, and resistance against predators and abiotic stress factors with special emphasis on the functions of indigenous microorganisms in the main ecological processes, e.g., in nutrient cycling.

It is obvious that only a few of these tasks can be solved by geneticists alone. There is an urgent need for a close cooperation between geneticists microbial ecologist, soil scientists, hydrogeologists, and environmental engineers in order to make genetically engineered microorganisms useful instruments in environmental biotechnology, and to prevent harm to man and the natural resources such as soil and groundwater.

References

1. Panel Discussion: Vulnerability of Ecosystems. In Omen, G.S. and Hollaender, A. (eds.) (1984), Genetic Control of Environmental Pollutants. Plenum, New York and London, pp. 169-186.
2. Filip, Z. (1975). Wechselbeziehungen zwischen Mikroorganismen und Tonmineralen und ihre Auswirkung auf die Bodendynamik. Habilitation-Thesis, Justus-Liebig-Universität Giessen, 172 pp
3. Hattori, T. (1973). Microbial Life in the Soil. Dekker, New York, 427 pp
4. Filip, Z. (1973). Fol. Microbiol. 18, 56-74
5. Fanning, D.S. and Keramidas, V.Z. (1977). In Dixon, J.B. and Weed, S.B. (eds.) Minerals in Soil Environments. Soil Sci. Soc. Am., Madison, pp. 195-258
5. Borchard, G.A. (1977). In Dixon, L.B. and Weed, S.B. (eds.) Minerals in Soil Environments. Soil Sci. Soc. Am., Madison, pp. 293-330
7. Filip, Z. (1970). Landbau Völ. 20, 91.96
8. Kanazawa, S. and Filip, Z. (1986). Microb. Ecol. 12, 205-210
9. Weinberg, S.R. and Stotzky, G. (1972). Soil Biol. Biochem. 4, 171-180
10. Lorenz, M.G. (1986). Ph.D.-Thesis, Universität Oldenburg
11. Lorenz, M.G., Aardema, B.W. and Krumbein, W.E. (1981). Mar. Biol. 64, 225-230
12. Zvyagintsev, D.G. (1973). Interactions between Microorganisms and Solid Surfaces. Izd. Moscow Univ., 175 pp (in Russian)
13. Filip, Z. and Hattori, T. (1984). In Marshall, K.C. (ed.) Microbial Adhesion and Aggregation. Springer-Verlag, Berlin, Heidelberg, New York, Tokyo, pp. 251-282
14. Roszak, D.B. and Colwell, R.R. (1987) Microb. Rev. 51, 365-379
15. Watter, M.V., Barbour, K., McDowel, M. and Seidler, R.J. (1987). Current Microbiol. 15, 193-197
16. Armstrong, J.L., Knudsen, G.R. and Seidler, R.J. (1987). Current Microbiol. 15, 229-232

17. Filip, Z., Dizer, H., Kaddu-Mulindwa, D., Kiper, M., Lopez--Pila, J.M., Milde, G. , Naser, A. and Seidel, K. (1986). Untersuchungen über das Verhalten pathogener und anderer Mikroorganismen und Viren im Grundwasser im Hinblick auf die Bemessung von Wasserschutzzonen. WaBoLu Hefte, Nr. 3, Bundesgesundheitsamt berlin, 121 pp

18. Filip, Z. (1979). Eur. J. Appl. Microbiol. Biotechnol. 6, 87-94

19. Stotzky, G., Devanas, M.A. and Zeph, L.R. (1988). In Filip, Z. (ed.) Biotechnologische 'in situ'-Sanierungsmaßnahmen im Boden- und Grundwasserbereich und ihre umwelthygienische Bedeutung. Fischer, Stuttgart (in print)

PLASMID TRANSFER IN SOIL AND RHIZOSPHERE

Jan Dirk van Elsas*, Jack T. Trevors** and Mary-Ellen Starodub**.

*Research Institute Ital, P.O.Box 48, 6700 AA Wageningen, The Netherlands

**Department of Environmental Biology, University of Guelph, Guelph, Ontario, Canada N1G 2W1

Summary: Transfer of plasmids pFT30 and RP4 between newly-introduced bacilli and pseudomonads, respectively, is described. In sterile soil, transfer of pFT30 was only detected when exogenous nutrients were added; bentonite clay significantly enhanced the transfer frequency. In nonsterile soil, pFT30 was transferred in the presence but not in the absence of bentonite clay. No transfer was detected in the rhizosphere of wheat. In sterile soil, transfer of RP4 between introduced pseudomonads was detected, and nutrients and bentonite clay both enhanced the transfer frequency. In nonsterile soil, transfer only occurred if exogenous nutrients were added; the transfer frequency was enhanced in bentonite-amended soil. In nutrient-enriched soil, transfer was detected at 15, 20 and 27 °C, but not at 4 and 10 °C. The plasmid transfer frequency was further significantly enhanced in the wheat rhizosphere as compared to transfer in bulk soil.

Key words: plasmid transfer, soil, rhizosphere, bacilli, pseudomonads.

Introduction

Naturally-occurring gene transfer has been identified as a potential risk factor which should be taken into consideration when genetically-engineered micro-organisms are to be released in the environment. In particular, this is so in those organisms in which the heterologous DNA is carried on (self-transmissable or mobilizable) plasmids, but even chromosomally-located heterologous DNA might eventually be picked up and transferred to other soil microbes by incoming plasmids.

Since the initial detection, by Weinberg and Stotzky (1) of the conjugal transfer of genetic material in sterile soil, other reports have described this process in sterile soil (2, 3) or, very recently, in nonsterile soil (2, 4). However, these studies were concerned with introduced strains of *Escherichia coli*, an organism considered alien to soil. Hence, the data

Risk Assessment for Deliberate Releases
Edited by W. Klingmüller
© Springer-Verlag Berlin Heidelberg 1988

obtained have limited predictive value for phenomena that may play a role when typical soil organisms with potential for agricultural applications, such as *Bacillus* spp. and *Pseudomonas* spp. (5) are introduced.

Although a few studies have shown the transfer of genetic material between plant-associated bacteria in the plant rhizosphere (6, 7) or *in planta* (8, 9), information on the influence of the plant on *in situ* gene transfer is still scarce.

Recently, in our laboratory the conjugal transfer of a plasmid encoding tetracycline resistance, pFT30, between two soil-isolated bacilli was demonstrated in soil (10). We now extend these results with studies on the transfer of pFT30 in the wheat rhizosphere. Also, an analysis of the *in situ* transfer of another plasmid, RP4, within a system consisting of two grass rhizosphere-isolated pseudomonads is reported.

Materials and methods

Organisms and plasmids

The bacterial strains and plasmids used are listed in Table 1. Strains of *Bacillus* were maintained and cultured using tryptone-yeast extract (TY) solid and liquid media, as described previously (10). For culturing and selecting *B. cereus* FoTc-30 and other strains containing plasmid pFT30, 10 µg/ml tetracycline (Tc) was used, whereas *B. subtilis* SEm-2 was grown using 10 µg/ml erythromycin (Em). Transconjugants were selected using both antibiotics. Plasmid pFT30 is considered to be related to reference plasmid pBC16 on the basis of its molecular weight (2.8 Mdal), tetracycline resistance determinant and stability in *B. cereus* and *B. subtilis* (11). In addition, plasmid DNA restriction patterns, obtained as described (12), were identical between pFT30 and pBC16 (Table 1), and the number and sizes of the fragments were in agreement with published values (13).

Pseudomonas sp strain R2f was isolated from grassland soil in the Netherlands. It was identified as a member of the fluorescent pseudomonad group using results from API 20E and API 50CH test strips. In addition, it was gram-negative, oxidase-positive and motile, and produced a fluorescent pigment on King's medium B (KB). To obtain a plasmid donor strain, plasmid RP4 was introduced into R2f by the filter mating technique (14) using *Escherichia coli* PC2366(RP4) as a donor. Plasmid RP4 is self-transmissable, encodes resistance to kanamycin (Km), Tc and carbenicillin, and is relatively

Table 1. Bacterial strains and plasmids

Strains	Antibiotic resistance*	Source/reference
Bacillus cereus FoTc-30	$Tc^r Pc^r$	(11)
Bacillus subtilis SEm-2	$Em^r Sm^r$	(11)
Bacillus subtilis 168 (pBC16)	Tc^r	(11)
Bacillus subtilis 168 (pFT30)	Tc^r	(10)
Escherichia coli PC2366 (RP4)		Phabagen Coll., Utrecht
Pseudomonas sp R2f		Dutch grassland soil
Pseudomonas sp R2f(RP4)	$Km^r Tc^r Cb^r$	This paper
Pseudomonas sp R2fRp^{r}Sm^r	$Rp^r Sm^r$	This paper
Plasmids	**Characteristics****	**Source/reference**
pFT30, pBC 16 ***	Tc^r; 2.8 Mdal; E 2, B 1, HI 1, HII 2, P 3, X 1, S 9-10, T 6	(11, this paper)
RP4	Km^r, Tc^r, Cb^r; 36 Mdal	(15)

* Tc=tetracycline, Pc=penicillin, Em=erythromycin, Sm=streptomycin, Km=kanamycine, Cb=carbenicillin, Rp=rifampicin.

** Sequentially: Antibiotic resistance; Molecular weight; Number of restriction sites for each enzyme used (E=EcoRl, B=BamHl, HI=HpaI, HII=HpaII, P=PvuII, X=XbaI, S=Sau3Al, T=Taql), e.g. E 2 signifies 2 sites for EcoRl.

*** pFT30 and pBC16 are identical according to the criteria used.

large (Mol.Wt 36 Mdal) (15). A spontaneous rifampicin (Rp) and streptomycin (Sm)-resistant mutant was selected to serve as a recipient strain in soil matings. All pseudomonads were maintained at -80 ^oC in 20 % glycerol. Donor strain R2f (RP4) was cultured and selected using liquid or solid KB media, supplemented with 75 µg/ml Km and 75 µg/ml Tc. Recipient strain R2f Rp^r Sm^r was cultured in KB + 50 µg/ml Rp + 50 µg/ml Sm, and selected on KB + 50 µg/ml Rp. Putative transconjugants were selected on KB + Km + Tc + Rp, and individual fluorescent colonies were checked for resistance to Sm. Plasmid RP4 was readily transferred in filter matings, between donor and recipient strains, at a maximal frequency per donor of roughly 20 %. In all experiments using nonsterile soil, 100 µg/ml cycloheximide (CH) was included in the media as an antifungal agent.

Plasmid detection

Detection of plasmid pFT30 in bacilli was as described (10). Plasmid RP4 was extracted from pseudomonads using the alkaline sodium dodecyl sulphate method and detected by agarose gel electrophoresis as described (18). Eight randomly selected putative RP4 transconjugants, from bulk and rhizosphere samples, revealed the presence of a plasmid with molecular weight equal to RP4, suggesting they were authentic transconjugants.

Soil and soil preparation

The soil used, Ede loamy sand, is a moderately acid "beekeerd" soil. Its characteristics have been described earlier (16). Freshly collected soil was used for each set of transfer experiments. In the laboratory, soil was sieved (4 mm mesh) and stored at room temperature at 15 % moisture in closed plastic bags. Where appropriate, soil was sterilized by γ-irradiation (4 MRad). Bentonite clay amendments were made as described (10). Before the onset of each experiment, soil was dried at room temperature to 8-10 % moisture.

Soil experiments

Transfer studies in bulk soil were carried out in 15 g of soil packed in 15 cm glass tubes (internal diameter 22 mm) to a bulk density of approximately 1.35 g/cm^3. The soil was homogeneously inoculated with 2 x 10^7-10^8 donor and recipient cells per g, adjusting the moisture content to about 20 % before introduction into the tubes. Cell suspensions used for nutrient-enriched soil conditions were prepared as described (10), whereas suspensions for nutrient-less conditions were in saline. After incubation, the soil from the tubes was analysed for the numbers of donor, recipient and transconjugant cfu by preparing a soil suspension as described (10) and subsequent dilution plating on the appropriate selective media.

To study the influence of wheat (*Triticum aestivum* var *Sicco*) roots on the transfer of plasmid pFT30, the soil chamber described by Dijkstra *et al.* (17) was employed. A modification of this chamber, consisting in the omission of the membrane, thus permitting free growth of roots into the soil, was used to study transfer of plasmid RP4. Soil was homogeneously inoculated with 2 x 10^7 to 10^8 donor and recipient cells (in saline) per g, to a final moisture level of 16 to 20 %. The soil was then packed in the soil chambers establishing a bulk density of approximately 1.35 g/cm^3. Wheat seeds were germinated on moist filter paper in Petri dishes for 3 to 4 days at room temperature. These wheat seedlings were placed on the membrane or soil surface (modified method) and a thin layer of gravel was applied on top. The soil chambers were placed on a moisture tension table applying a water tension of 100 cm according to Dijkstra et al. (17), and incubated in a 20 °C climate room with dark/light regime (16/8 h).

In the original set-up, the membrane plus adhering soil was removed from the soil after incubation; this was considered the rhizosphere sample. Further, two soil slices were made (17), without freezing the soil, establishing bulk soil samples. In the modified method, the wheat plants plus the tightly adhering soil were carefully taken out of the soil, overground parts were cut off, and the resulting sample was considered the rhizosphere sample. A 10 g portion of the remaining soil (bulk soil sample) was also obtained. All samples were processed as indicated above for bulk soil samples.

The counts from all experiments were expressed on a soil dry weight base. The plasmid transfer frequency was calculated as the number of transconjugant cfu/number of initially introduced donor cfu. All experiments were carried out in duplicate.

Results and discussion

Transfer of plasmid pFT30 between bacilli in soil and rhizosphere

Van Elsas *et al*. recently demonstrated the transfer of plasmid pFT30 in soil from *Bacillus cereus* FoTc-30 to *B. subtilis* SEm-2 at 27 °C and 20 to 22% soil moisture (10). From *in vitro* experiments, it was inferred that the mechamism of transfer was probably conjugation. In sterile soil, plasmid transfer only occurred in the presence of added, easily-degradable nutrients. Similar observations have recently been made in studies with *E. coli* transfer systems in soil (3). The presence of bentonite clay in the soil further significantly enhanced the transfer rate. At 15 °C or at 8 % soil moisture, plasmid transfer also occurred, albeit at a lower rate. In non-sterile soil, the introduced recipient population declined rapidly and no transconjugants were detected; however, in the presence of bentonite clay survival of the recipient population was significantly enhanced and plasmid transfer was observed. Protection of bacteria by clay minerals has been previously described (19, 20), and the enhanced survival of the recipient cells may have resulted in detectable plasmid transfers.

On the basis of the stimulating effect of easily-degradable compounds on plasmid transfer, it was decided to study this process in the rhizosphere of wheat using the Dijkstra soil chamber (17) packed with either bentonite-amended or unamended soil.

Table 2. Survival of donor and recipient strains and transfer of plasmid pFT30 in nonsterile bulk and rhizosphere soil*

Strains	Soil treatment	3 hours (Bulk)	7/10 days Rhizosphere (0-0.7 mm)	Bulk 1 (0.7-2.7 mm)	Bulk 2 (2.7-6.0 mm)
Donors	-B**	9.4×10^6	7.9×10^5	3.6×10^5	8.7×10^4
	+B	6.5×10^6	1.1×10^6	5.0×10^5	3.2×10^5
Recipients	-B	2.2×10^5	1.4×10^4	4.4×10^3	3.2×10^3
	+B	2.0×10^6	4.2×10^4	3.6×10^4	3.5×10^4
Transconj.	-B	ND	ND	ND	ND
	+B	ND	ND	ND	ND

* Numbers of cfu/g of dry soil after 3 hours and 7 (soil without bentonite) or 10 (soil with bentonite) days of soil incubation. Geometric means of duplicate samples. ND=not detected on plates prepared from undiluted soil suspensions.
** B=bentonite clay.

Results showed that both the donor and, more in particular, the recipient populations survived poorly in nonsterile soil (Table 2). The poor survival was apparently not counteracted by the presence of wheat roots, since the cell numbers detected in the rhizosphere were similar to or slightly higher than those in the bulk soil. In addition, the presence of bentonite clay only partially improved bacterial survival. No transconjugants were detected in any of the samples. This may have been due to the poor survival of the introduced populations. Poor survival of introduced recipient cells and lack of transfer of pFT30 in both bulk and rhizosphere was also previously found in sterile soil (not shown). However, in this experiment, the introduced donor population showed stable dynamics in the bulk soil and a slight stimulation in the rhizosphere.

Transfer of plasmid RP4 between pseudomonads in soil and rhizosphere

To simulate environmental conditions realistic for Dutch soils, initial experiments in sterile and non-sterile soil were carried out at 20 ^oC. The results are contained in Table 3. In sterile soil without added nutrients, both the introduced donor and recipient populations showed good survival. After 3 hours, low-frequency transfer of plasmid RP4 was detected, and frequency remained roughly stable upon further incubation. In the presence

Table 3. Transfer of plasmid RP4 between pseudomonads after 3 and 24 hours in sterile and nonsterile soil, at 20 °C, under different conditions*.

Soil conditions**	Time (hours)	No. of donor cells	No. of recipient cells	No. of transconj.	Transfer freq.***
Sterile,-nutrients	3	2.4×10^{8}	2.75×10^{8}	8.6×10^{2}	
	24	1.9×10^{8}	5.5×10^{7}	1.2×10^{3}	1.2×10^{-5}
Sterile,+nutrients	3	8.6×10^{7}	1.0×10^{8}	3.1×10^{3}	
	24	3.5×10^{7}	4.7×10^{8}	2.8×10^{4}	2.8×10^{-4}
Sterile,-nutr.,+B	3	2.6×10^{8}	2.2×10^{8}	2.9×10^{4}	
	24	5.8×10^{8}	5.6×10^{8}	1.4×10^{5}	1.4×10^{-3}
Sterile,+nutr.,+B	3	8.4×10^{7}	1.2×10^{8}	5.9×10^{4}	
	24	2.0×10^{9}	1.2×10^{9}	2.6×10^{6}	2.6×10^{-2}
Nonster.,+nutr.,-B	3	4.1×10^{6}	1.3×10^{6}	1.25×10^{2}	
	24	1.3×10^{6}	1.5×10^{7}	4.4×10^{3}	4.4×10^{-5}
Nonster.,+nutr.,+B	3	5.7×10^{7}	7.3×10^{7}	3.2×10^{4}	
	24	6.0×10^{8}	1.2×10^{9}	1.6×10^{6}	1.6×10^{-2}

* Numbers of cfu per g of dry soil. Results are geometric means of duplicate samples.
** B=bentonite clay.
*** After 24 h.

of added nutrients, the introduced recipient population proliferated, whereas the plasmid-containing donor cell population declined. In addition, the plasmid transfer frequency was significantly enhanced as compared to soil without nutrients ($P<0.05$). The decline in the plasmid-containing donor cell population detectable on selective plates may be explained by assuming that plasmid loss occurred during growth in the sterile soil. The total cell count (6.6×10^{8} cfu/g of dry soil), slightly higher than the donor plus recipient cell count, also suggested the presence of plasmid-less donor cells. Loss of plasmid RP4 in soil was previously detected by Schilf and Klingmüller (21), and in pure culture by Saunders and Grinsted (15).

In bentonite-amended sterile soil, both in the absence and presence of added nutrients the survival of donor and recipient populations was improved as compared to the survival in unamended soil in both conditions (Table 3). In addition, the plasmid transfer frequency was significantly enhanced in both conditions ($P<0.05$). In nonsterile unamended soil in the presence of added nutrients, the introduced donor and recipient populations survived poorly, and only low-frequency transfer was recorded. Competition, antagonism or antibiosis by the indigenous microflora may have caused the poor survival

Table 4. Transfer of plasmid RP4 between pseudomonads after 24 hours in nutrient-enriched and unenriched nonsterile soil at different temperatures*.

Soil temperature	No. of donor cells	No. of recipient cells	No. of trans-conjugants	Transfer frequency
4 °C (+nutrients)	$<10^5$	5.7×10^6	ND	ND
10 °C	1.2×10^5	1.6×10^6	ND	ND
15 °C	4.2×10^5	9.4×10^7	65	3.3×10^{-6}
20 °C	2.9×10^6	4.3×10^7	8.3×10^2	4.2×10^{-5}
27 °C	1×10^5	7.4×10^5	5.1×10^2	2.6×10^{-5}
27 °C (-nutrients)	1.1×10^6	2.7×10^6	ND	ND

*Numbers of cfu per g of dry soil. Results are geometric means of duplicate samples. ND=not detected on plates prepared from undiluted soil suspensions.

and low transfer frequency. Moreover, in previous separate experiments in nonsterile soil, no plasmid transfer was detected in the absence of added nutrients. This suggested that transfer was probably stimulated by the presence of nutrients but that this effect was counteracted by the concomitant stimulation of deleterious biotic effects. In nutrient-enriched nonsterile soil amended with bentonite clay, both the introduced donor and recipient populations revealed a significantly better survival ($P<0.05$) than without bentonite clay and the transfer frequency was also significantly enhanced ($P<0.05$). This again suggested that bentonite clay is able to protect introduced bacteria from deleterious soil effects, thereby also stimulating plasmid transfer. This protection was almost complete, since similar numbers of donor, recipient and transconjugant cells were detected in the bentonite-amended sterile and nonsterile soils (Table 3).

In the second set of experiments, the influence of temperature and added nutrients was studied in nonsterile soil (Table 4). Both at 4 and 10 °C, poor donor and recipient survival was noted in nutrient-enriched soil, and no plasmid transfer was detected. Survival of the introduced donor population was also poor at 15, 20 and 27 °C, whereas the recipient population survived well at 15 and 20 °C, but poorly at 27 °C. This differential behaviour of the introduced populations at different temperatures may have been caused by shifts in the indigenous microflora or by different sensitivity to soil conditions induced by temperature changes. At 15, 20 and 27 °C, low-frequency plasmid transfer occurred in the nutrient-enriched soil (Table 4). The detection of transfer at 15 °C is consistent with recent data on plasmid transfer between introduced E. coli strains at this temperature (3). Apparently, the previous contention that mating pair formation does not occur below 24 °C (22) does not hold strictly for matings between bacteria in soil.

At 27 °C in the absence of added nutrients, the introduced donor and recipient populations again declined rapidly and no plasmid transfer was found, similarly to what was previously found at 20 °C. This observation is in agreement with previous findings on plasmid transfer between E. coli strains in sterile soil systems (3).

To investigate whether the presence of growing wheat roots affected plasmid transfer between the two strains distributed homogeneously in the soil, the modified soil chamber of Dijkstra et al. (17) was used. The results (Table 5) indicated that no pronounced rhizosphere effect occurred for either one of the introduced strains, while both on days 5 and 10 a significant enhancement ($P<0.05$) of transfer frequency occurred in the rhizosphere soil as compared to the bulk soil. The enhanced plasmid transfer frequency in the rhizosphere may be due to an increase in transfers, caused either by higher metabolic activities of the introduced cells induced by root exudates, or by increased possibilities for cell-to-cell contact at the root surfaces. An alternative explanation may be better survival or growth of transconjugants in the rhizosphere. Experiments aimed at studying the nature of the effect, as well as how far the effect extends into the soil, will be published elsewhere.

Table 5. Transfer of plasmid RP4 between pseudomonads in bulk and rhizosphere soil*.

	3 hours	5 days		10 days	
	(Bulk)	Bulk	Rhizosphere	Bulk	Rhizosphere
Donors	1.2×10^{8}	2.7×10^{7}	2.1×10^{7}	4.0×10^{6}	9.4×10^{6}
Recipients	7.4×10^{8}	1.0×10^{8}	5.7×10^{7}	5.2×10^{6}	1.3×10^{7}
Transconjugants	3.7×10^{3}	1.2×10^{2}	6.5×10^{4}	ND	6.0×10^{4}
Transfer freq.	2.9×10^{-5}	0.9×10^{-6}	5.0×10^{-4}	ND	4.6×10^{-4}

*Numbers of cfu per g of dry soil. Results are geometric means of duplicate samples. ND=not detected on plates prepared from undiluted soil suspensions.

In conclusion, the occurrence of (conjugal) plasmid transfer in soil between introduced bacilli or pseudomonads is related to the survival of sufficient numbers of introduced cells, since at higher cell densities chances of cell-to-cell contact evidently increase. In addition, the characteristics of the bacteria and plasmids used also determine the outcome of transfer studies in soil. The enhanced plasmid transfer in the presence of both nutrients and bentonite clay may be largely due to effects on bacterial activity and survival. The stimulation of genetic transfer between pseudomonads in the rhizosphere of wheat, and the absence of transfer between the bacilli were in line with the suggested competitive ability of pseudomonads in the rhizosphere and the relative rhizosphere incompetence of bacilli such as Bacillus subtilis (23). Since the utilization of genetically-engineered bacteria for agricultural purposes is often cogitated, the results here presented stress the necessity to assess the risk of horizontal gene transfer in different soil conditions, in particular in the rhizosphere.

Acknowledgements

Excellent technical assistance by J.M. Govaert and L. van Overbeek is acknowledged. We are further grateful to Phabagen Collection, Utrecht, for submitting an Escherichia coli plasmid donor strain. Appreciation is further expressed to J.H. Oude Voshaar for statistical help, and L. van Vloten-Doting and J.A. van Veen for reading the manuscript. This work was supported by a NATO Collaborative grant awarded to J.D.V.E. and J.T.T.

References

1. Weinberg, S.R. and Stotzky, G. (1972). Soil Biol. Biochem. 4, 171-180.
2. Krasovsky, V.N. and Stotzky, G. (1987). Soil Biol. Biochem. 19, 631-638.
3. Trevors, J.T. and Oddie, K.M. (1986). Can. J. Microbiol. 32, 610-613.
4. Trevors, J.T. and Starodub, M.E. (1987). System. Appl. Microbiol. 9, 312-315.
5. Burr, T.J. and Caesar, A. (1984). Crit. Rev. Plant Sci. 2, 1-20.
6. Plazinski, J. and Rolfe, B.G. (1985). Con. J. Microbiol. 31, 1026-1030.
7. Talbot, H.W., Yamamoto, D.K., Smith, M.W. and Seidler, R.J. (1980). Appl. Environm. Microbiol. 39, 97-104.
8. Johnston, A.W.B. and Beringer, J.E. (1975). J. Gen. Microbiol. 87, 343-350.
9. Manceau, C., Gardan, L. and Devaux, M. (1986). Can. J. Microbiol. 32, 835-841.

10. Van Elsas, J.D., Govaert, J.M. and Van Veen, J.A. (1987). Soil Biol. Biochem. 19, 639-647.

11. Van Elsas, J.D. and Pereira, M.T.P.R.R. (1986). Plant Soil 94,213-226.

12. Maniatis, T., Fritsch, E.F. and Sambrook, J. (1982). Cold Spring Harbor Lab.

13. Perkins, J.B. and Youngman, P. (1983). J. Bacteriol. 155, 607-615.

14. Simon, R., Priefer, U. and Pühler, A. (1983). Biotechnol. 1, 784-791.

15. Saunders, J.R. and Grinsted, J. (1972). J. Bacteriol. 112, 690-696.

16. Van Elsas, J.D., Dijkstra, A.F., Govaert, J.M. and Van Veen, J.A. (1986). FEMS Microbiol. Ecol. 38, 151-160.

17. Dijkstra, A.F., Govaert, J.M., Scholten, G.H.N. and Van Elsas, J.D/. (1987). Soil Biol. Biochem. 19, 351-352.

18. Trevors, J.T. (1987). Water, Air Soil Pollut. 34, 409-414.

19. Van Veen, J.A. and Van Elsas, J.D. (1987). In: Proc. IV Int. Symp. Microb. Ecol. (Ljubliana), in press.

20. Stotzky, G. (1986). In: Interactions of Soil Minerals with Natural Organics and Microbes (Eds. Huang, P.M. and Schnitzer, M.), pp. 305-428.

21. Schilf, W and Klingmüller, W. (1983). Recomb. DNA Techn. Bull. 6, 101-102.

22. Wamsley, R.H. (1976). J. Bacteriol. 126, 222-224.

23. Dijkstra, A.F., Scholten, G.H.N. and Van Veen, J.A. (1987). Biol. Fertil. Soils 4, 41-46.

BACTERIA WITH NEW PATHWAYS FOR THE DEGRADATION OF POLLUTANTS AND THEIR FATE IN MODEL ECOSYSTEMS

D.F. Dwyer, F. Rojo and K.N. Timmis
Dept. de Biochimie Medicale, Université de Genève
9, avenue de Champel, CH-1211 Genève 4

Summary: Two genetically engineered bacteria were constructed by a patchwork assembly of cloned genes from different bacteria into *Pseudomonas* sp. strain B13. The engineered bacteria were able to degrade chlorinated and methylated benzoic acids which often exist simultaneously in industrial waste streams at concentrations which inhibit their degradation by indigenous microorganisms. The bacterial strains were shown to survive within a model aerobic sludge ecosystem and were able to degrade substituted benzoates that were present in ordinarily inhibitory combinations.

Introduction

The potential for using genetically engineered microorganisms (GEMs) to degrade environmental pollutants and to cleanse industrial and municipal waste streams has yet to be realized, but promises to become increasingly important as a means for pollution control, especially when one considers the large quantities of biodegradatively recalcitrant and often toxic pollutants in the environment (1). Three factors presently restrict the engineering and use of pollutant-degrading bacterial strains: (i) A large amount of effort is required in identifying and cloning genes for the degradation of recalcitrant organic compounds. Following this, there must be a logical construction of suitable catabolic pathways which are regulated for efficient activity (2,3). (ii) Little information is currently available for predicting the ability of GEMs to survive and function *in vivo*. A need exists for creating and testing model GEMs to determine their response to given environmental factors (4). (iii) The safety of

Risk Assessment for Deliberate Releases
Edited by W. Klingmüller
© Springer-Verlag Berlin Heidelberg 1988

introducing GEMs into the environment, either deliberately or accidentally is of concern. The use of model ecosystems is one approach to determine the effect of the GEM on its target and related environments before introduction is attempted (5,6).

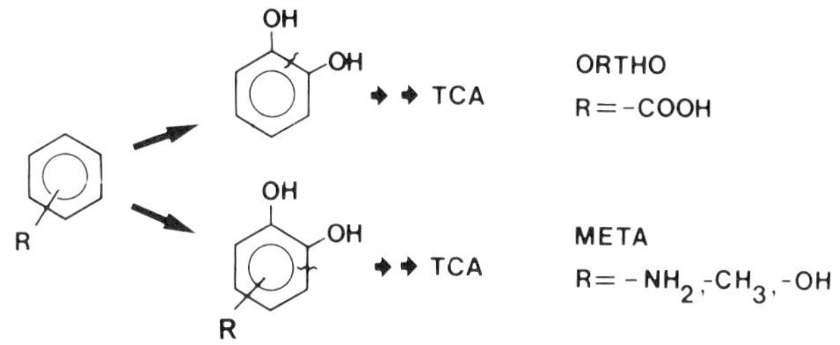

Figure 1 Benzoate is metabolized predominantly through an ortho-cleavage pathway while alkyl-aromatics are degraded through the meta-cleavage pathway. In nature, other substituted aromatics also are predominantly metabolized through the meta-cleavage pathway. The site of ring cleavage is demonstrated.

We have attempted to sequentially address these three issues by constructing pollutant-degrading GEMs and testing their response to in vivo conditions manifested within a model aerobic sludge ecosystem. The bacteria were designed to degrade methyl- and chloro-substituted aromatic compounds. A good deal of information is known about the degradative routes of these compounds (7) which can be metabolized either through an intradiol- (ortho-) or an extradiol- (meta-) cleavage pathway (Fig. 1). In soil and sewage ecosystems, benzoate is channeled through the ortho-cleavage route. Aromatics with substitutents other than carboxyl groups, eg. alkyls such as methyl groups, are channeled through the meta-cleavage pathway. Bacteria often possess both types of pathways. When a halogen substituent is present, degradation results in the production of halocatechols. These compounds are metabolized inefficiently by the ortho-cleavage route and act as inactivators of the 2,3-dioxygenase of the meta-cleavage route. When both alkyl-substituted and halogen-substituted aromatics are present simultaneously, both the meta- and ortho-cleavage routes are induced, resulting in the routing of substituted catechols into unproductive pathways and the formation of recalcitrant halocatechols.

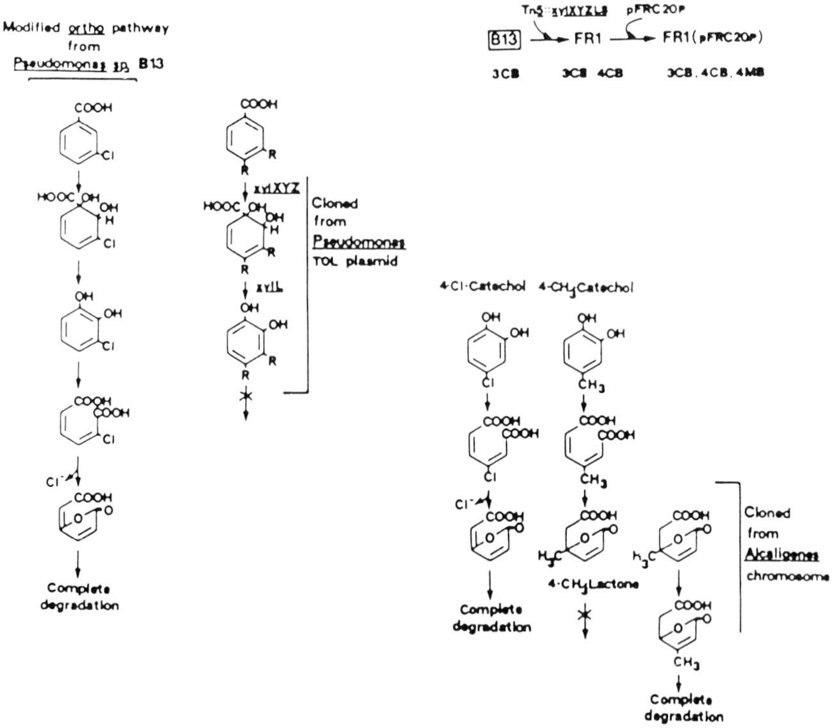

Figure 2 The ortho pathway for the degradation of 3-chlorobenzoic acid (3CB) of Pseudomonas sp. strain B13 was modified by the introduction of the TOL plasmid genes coding toluate 1,2-dioxygenase (xylXYZ) and dihydroxycyclohexadiene carboxylate dehydrogenase (xylL) which expanded the degradation range to include 4CB and permitted transformation of 4-methylbenzoate to 4-methyl-2-enelactone. 4-methyl-2-enelactone isomerase was recruited from Alcaligenes eutrophus, this allowed transformation of 4-methyl-2-enelactone to 3-methyl-2-enelactone which is degraded by B13. The genetic steps leading to new strain formation are outlined in the upper right corner.

Construction of genetically engineered strains

One solution for this problem is to construct within the same microorganism a catabolic sequence for the degradation of both methyl-aromatics and halo-aromatics through an ortho-cleavage route. This required the patchwork assembly of genes encoding the necessary enzymes from three different bacteria (Fig. 2). Pseudomonas sp. strain B13 was used as the organism in which the pathway was assembled because it contains a well-characterized ortho-cleavage pathway for the degradation of 3-chlorobenzoic acid (3CB) and lacks detectable meta-cleavage activity (9). Recruitment into strain B13 of the TOL plasmid-encoded enzymes toluate dioxygenase and dihydrocyclohexadiene carboxylate dehydrogenase allowed the new hybrid to degrade 4-chlorobenzoic acid (4CB) and transform methyl-benzoates via methylcatechols to 4-carboxymethyl-methylbut-2-ene-1,4-olide (4-methyl-2-enelactone). This GEM was designated Pseudomonas sp. strain FR1 (FR1) and was kanamycin-resistant (Km^r).

To complete metabolism of 4-methyllactone, the enzyme 4-methyl-2-enelactone isomerase was recruited from Alcaligenes eutrophus into FR1. This allowed the hybrid strain designated Pseudomonas FR1(pFRC20P) to convert 4-methyl-lactone to 3-methyl-lactone which could then be completely metabolized (Fig. 2). Thus, by using in vitro methods of genetic manipulation, a strain of bacterium was constructed which could simultaneously degrade 3CB, 4CB and 4MB and which had potential use for the cleansing of waste streams which contain mixtures of these compounds. Both constructed strains, FR1 and FR1(pFRC20P), were then tested in model ecosystems to determine their ability to survive and function under simulated in vivo conditions.

Model ecosystems and bacterial enumeration

The model aerobic sludge ecosystem was maintained in a 2.5 liter bottle and consisted of 900 ml of synthetic sewage (OECD, Anon., 1971) plus an initial inoculum of 100 ml of activated sludge obtained from a municipal water purification plant. Air was vigorously pumped into the bottles; continuous dilution with fresh synthetic sludge amended with different substituted benzoates was provided at a dilution rate (D) of 0.05/h. Pseudomonas sp. strains B13, FR1 or FR1(pFRC20P) were added to the ecosystem at levels approximating 1% of the total bacterial population. Total colony forming units (CFU) were enumerated daily; B13 and derivative strains were enumerated on selective media (Figures 3 and 4). The presence of a meta-cleavage route of degradation of aromatics was demonstrated by spraying the agar plates with a solution of catechol which caused colonies containing the meta-cleavage enzyme catechol 2,3-dioxygenase to turn yellow upon conversion of catechol to 2-hydroxy muconic semialdehyde.

Survival and functioning of bacterial strains in the ecosystems

Pseudomonas sp. strain B13 initially was added to the ecosystems at approximately 10^7 bacteria/ml. This number declined to a level of about 1.0×10^5 bacteria/ml (Fig. 3A). The total number of CFUs demonstrated the typical oscillating pattern of a predator-prey relationship while the system equilibrated. This pattern was evident in several experimental trials and was similar to that observed with the two derivative strains, FR1 (Fig. 3) and FR1(pFRC20P) (Fig. 4). The synthetic medium contained 1 mM 4CB which was neither degraded by nor had any apparent effect on the survival of B13. A population of 4CB-degrading bacteria developed by day 10 at 1 x 102 bacteria/ml which increased to 2×10^5 bacteria/ml by day 14.

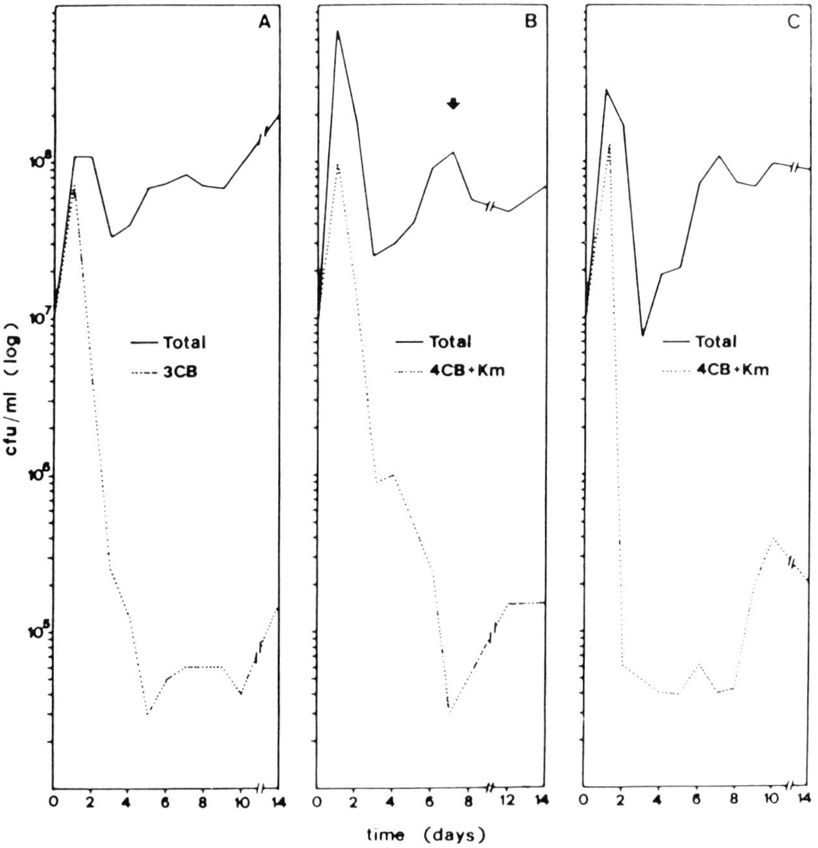

Figure 3 The model ecosystem for <u>Pseudomonas</u> sp. strain B13 (A) was provided with 1 mM 4CB from the onset; FR1 was provided with 1 mM 4CB either on day 8 (B,🡇) or continuously (C). FR1 was enumerated on agar plates containing 4CB plus kanamycin; B13 was enumerated on agar plates containing 3CB.

The genetic alterations which were used to construct FR1 apparently did not alter its ability to establish itself in the ecosystem from that of B13 (Fig. 3B and 3C). The GEM functioned as a 4CB-degrader as indicated by the degradation of 4CB (data not shown) and by the apparent lack of emergence of the new population of 4CB-degrading bacteria that had occurred in the ecosystem with <u>Pseudomonas</u> sp. strain B13 (Fig. 3A).

The presence of pFRC20P in FR1 did not alter its ability to establish itself in the ecosystem (Fig. 4A) and allowed

FR1(pFRC20P) to degrade combinations of 4MB and 3CB: (i) With continuous addition of 5 mM 4 MB and a shock load of 5 mM 3CB on day 3 which continued through day 12 (Fig. 2B), the 4 MB was degraded, but the shock load of 3CB was not (data not shown).

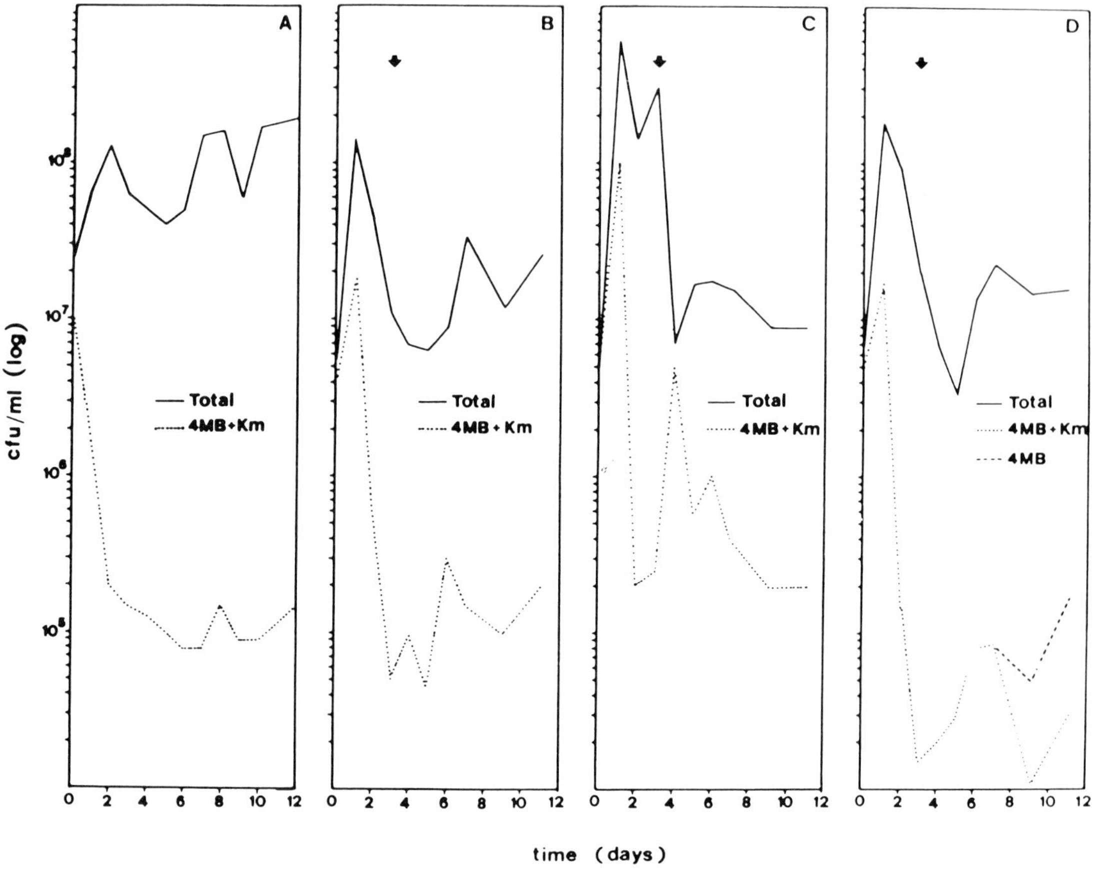

Figure 4 FR1(pFRC20P) was provided with synthetic sludge amended with 5 mM 3CB plus a shock load (🡇) of 5 mM 4MB from day 3 through day 12 (B), with 5 mM 4 MB plus a shock load (🡇) of 5 mM 3CB from day 3 through day 12 (C), and a simultaneous shock load of 5 mM 3CB and 5 mM 4MB from day 3 through day 12 (D). FR1(pFRC20P) was enumerated on agar plates containing 4MB plus kanamycin. The total population of 4MB-degrading bacteria was enumerated on agar plates containing 4MB alone.

(ii) The addition of substituted benzoates was reversed such that 5 mM 3CB was fed to the ecosystrem with a shock load of 5 mM 4MB from day 3 through day 12 (Fig. 4C). In this case the 3CB was substantially degraded while 4 MB accumulated (data not shown). In both shock load scenarios the concentration of the initial substituted benzoate was kept low enough that the subsequent addition of 3CB or 4MB had only a transitory effect on the ecosystem as evidenced by the decline of the total number of CFUs. (iii) When the shock load consisted of simultaneous additions of 5 mM 3CB and 5 mM 4MB from day 3 through day 12 (Fig. 4D), the effect on the ecosystem was more drastic and the number of CFUs declined to 2.5×10^6 bacteria/ml. In this case, an indigenous population of 4MB degrading bacteria with an ortho-cleavage route of degradation for 4MB developed in response to the additions.

Discussion

There are five basic factors which need to be considered before introduction of a GEM into the environment occurs: (i) whether the GEM will survive in the target environment, (ii) whether cloned genetic material will be stable within the GEM, (iii) what the potential is for transfer of cloned genetic material to indigenous microorganisms, (iv) whether the GEM will be able to efficiently function in the target environment in the task for which it was designed, and (v) what effects, if any, the GEM may have upon the ecosystem into which it is introduced or to which it may spread.

Short of actual direct introduction, the only way to study these factors is by model ecosystem studies. These model ecosystems should closely approximate the conditions of the target environment. Thus, if a GEM, however well conceived, either fails to survive or function in the model ecosystem or has an adverse effect, these traits may be discerned, thereby alleviating the need of more costly and time-consumming introduction experiments. The model ecosystem which we used in these experiments is simple in design, but closely simulated aerobic sludge in the diversity and number of microorganisms present and in nutrient conditions. The microorganisms included both predators of bacteria and bacteria able to successfully compete with the GEMs, such as those which degraded 4MB and were able to displace FR1(pFRC20P) (Fig. 4D).

A comparison of the response of parental and derivative GEMs in ecosystem studies can be used to give an indication of any effect(s) of genetic alteration upon the GEM. The similar patterns of response for B13, FR1 and FR1(pFRC20P), demonstrated that the effect of engineering did not adversely affect the GEM. The ability of the two GEMs to degrade substituted benzoates under simulated ecosystem conditions also demonstrated that a logical reassembly of the genes coding catabolic enzymes can be used to achieve the desired catabolic effects. Although neither strain completely degraded the added benzoic acids, FR1(pFRC20P) provided sufficient degradative capacity to buffer the ecosystem from shock loads of substituted benzoates. In pure culture with only the benzoates as substrates, FR1(pFRC20P) is able to totally degrade combinations of 3CB and 4MB, but probably failed to do so in this example because of the availability of prefered growth nutrients.

Both the stability of cloned genetic information in GEMs and the potential for its transfer to indigenous microorganisms can also be predicted by the use of model ecosystem. We are attempting to determine if transfer of the cloned genes to the indigenous microbial population and to suitable recipient bacteria added simultaneously to the model ecosystem can occur by screening colonies which grow upon selective medium with radioactively labeled probes for the cloned genes. As would be predicted, the potential for transfer of cloned genes is high when the genes are present on a mobilizable or conjugative plasmid and low when they are stably inserted into the chromosome of the parental strain.

References

1. Keith, L.H. and Telliard, W.A. (1979). Environ. Sci. Technol. 13, 416-423

2. Ramos, J.L., Stolz, A., Reineke, W. and Timmis, K.N. (1986). Proc. Natl. Acad. Sci. USA. 83, 8467-8471

3. Ramos, J.L. and Timmis, K.N. (1987). Microbiol. Sci. 4, 228-237

4. Levin, M.A., Seidler, R., Borquin, A.W., Fowle, J.R.III and Barkay, T. (1987). Bio/Technology 5, 38-45

5. Jain, R.K., Sayler, G.S., Wilson, J.T., Houston, L. and Pacia, D. (1987). Appl. Environ. Microbiol. 53, 996-1002

6. Liang, L.N., Sinclair, J.L., Mallory, L.M. and Alexander, M. (1982). Appl. Environ. Microbiol. 44, 708-714

7. Knackmuss, H.-J. (1983). in Biotechnology, eds. Phelps, C.F. and Clarke, P.H. (University Press, Cambridge), pp 173-190

8. Rojo, F., Pieper, D.H., Engesser, K.-H., Knackmuss, H.-J. and Timmis, K.N. (In press). Science

9. Reineke, W. and Knackmuss, H.-J. (1980). J. Bacteriol. 142, 467-473

IMPACT OF MINERAL SURFACES ON GENE TRANSFER BY TRANSFORMATION IN NATURAL BACTERIAL ENVIRONMENTS

Michael G. Lorenz and Wilfried Wackernagel
Arbeitsgruppe Genetik, Fachbereich Biologie
Universität Oldenburg, D-2900 Oldenburg, FRG

Summary: We have developed a microenvironmental system consisting of glass columns filled with quartz-rich chemically pure sea sand. The columns were percolated with medium. Using this system, we have determined parameters affecting adsorption and desorption of DNA to sand. Adsorbed DNA was shown to be more resistant to degradation by DNase I than free DNA. With DNA adsorbed to sand highly efficient homologeous transformation of *Bacillus subtilis* was obtained, exceeding transformation frequencies of standard liquid transformation by 25 to 50 fold. Furthermore, the accumulation of cells at sand grains resulted in up to 16% transformed cells at a temperature (23°C) suboptimal for transformation. It is also shown that the transformation process in sand is 10 fold more resistant to DNase I and 100 to 1000 fold more resistant to low temperatures (7°C) than transformation in liquid culture. The results show that persistance of DNA and gene transfer by genetic transformation are facilitated at the surface of mineral grains.

Introduction

Bacterial gene transfer in natural environments has recently become a subject of increasing experimental investigation (1). The present interest in these processes derives from the discussion whether or not released genetically engineered microorganisms may transfer foreign genes to the inhabiting microbial community. Also, there are experimental data and considerations suggesting a role of gene transfer and genetic recombination in the evolution of microorganisms (2, 3). Main efforts have so far concentrated on conjugation and less information is available on transduction and transformation in nature (1). In sterile soil transformation has been observed and was not inhibited by the addition of DNase (4).

Risk Assessment for Deliberate Releases
Edited by W. Klingmüller
© Springer-Verlag Berlin Heidelberg 1988

Table 1. Naturally transformable bacteria living in soil and sediment

Species	Transformation frequency (Recombinants/viable cells)	Reference
Photolithotrophic		
Synechocystis sp. 6803	5×10^{-4}	(8)
Synechocystis sp. OL50	2×10^{-4}	(9)
Agmenellum quadruplicatum	4.3×10^{-4}	(10)
Nostoc muscorum	1.2×10^{-3}	(11)
Chemolithotrophic		
Thiobacillus thioparus	10^{-3} to 10^{-2}	(12)
Thiobacillus sp. strain Y	1.7×10^{-3}	(12)
Heterotrophic		
Bacillus subtilis	3.5×10^{-2}	(13)
Acinetobacter calcoaceticus	7×10^{-3}	(14)
Azotobacter vinelandii	9.5×10^{-2}	(15)
Micrococcus radiodurans	2.1×10^{-2}	(16)
Pseudomonas stutzeri	7×10^{-5}	(17)
Mycobacterium smegmatis	10^{-7} to 10^{-6}	(18)
Thermoactinomyces vulgaris	2.7×10^{-3}	(19)
Methylotrophic		
Methylobacterium organophilum	5.3×10^{-3}	(20)
Archaebacteria		
Methanococcus voltae	8×10^{-6}	(21)

Natural habitats such as soil are characterized by organic and inorganic particulate material (5) and bacteria are mainly associated with these components (5, 6, 7). Many species of bacteria normally inhabiting soil and sediment have been reported to develop a natural state of competence (Tab. 1). Consequently, we have addressed in several lines of experiments the question of how the surface of mineral particulate material present in natural habitats may contribute to the probability and efficiency of gene transfer by genetic transformation. Our results obtained from experiments in a model microenvironment using *B. subtilis* as a typical ubiquitous bacterium reveal important functions of mineral surfaces for gene transfer by transformation: the facilitation of persistance of free DNA by adsorption and the enhancement of transformation. In this communication we briefly summarize some of our results (22, 23) and present some new data on the effect of temperature on the transformation process at the surface of sand grains.

Materials and methods

Bacteria: *Bacillus subtilis* 168 strain 1G20 (*trpC2*) was used for transformation throughout. The transforming DNA was prepared from a prototrophic *B. subtilis* 168 strain (DSM no. 401).

Microenvironment: Glass columns (5 x 70 mm; in temperature experiments glass jacketed) were filled with 0.7 g of analytical grade sea sand (Merck, Darmstadt) and autoclaved. The columns were percolated with buffered salt solution prior to DNA loading or with aerated medium after a sample of a competent culture had been applied to the sand. Elution was done using a calibrated high precision pump (Microperpex, LKB) at 0.2 ml/min (22).

Loading of columns with DNA: Calf thymus DNA (Boehringer, Mannheim) was used for adsorption and desorption experiments. DNA solutions (575 µg/ml) were dialysed against buffered salt solutions. Experiments on DNA degradation were done with transforming DNA (79 µg/ml) isolated from *B. subtilis*. Samples of DNA (0.2 ml) were applied to the column and non-adsorbed DNA was removed by elution with 5 ml of the loading buffer at 0.2 ml/min (22). Adsorbed calf thymus DNA was determined by the diphenylamine reaction (24) and adsorbed *B. subtilis* DNA by transformation.

Enzymatic degradation of DNA: Degradation of sand-adsorbed DNA was studied by applying to the column 0.2 ml of a DNase I solution (10 ng/ml in 50 mM $MgCl_2$, 10 mM Tris-HCl, pH 7.0) for 15 min at 23°C. The reaction was terminated with sodiumdodecylsulfate (22) and the remaining transforming activity of DNA was determined.

Transformation: Standard transformation in liquid medium was as described (25). Transformation on DNA-loaded sand grains was obtained by applying 0.2 ml of a competent culture (4 x 10^7 cells total) to a column which was then percolated with medium for 30 min. Termination of transformation by DNase I, recovery of cells and determination of transformants and viable counts was as described (23). In experiments without adsorbed DNA the cell-loaded column was eluted with medium for 30 min before 0.2 ml of DNA were added (23).

Results

Adsorption of DNA to sand:

The microenvironment used in this studies is shown in Fig. 1. Kinetic experiments revealed that adsorption was essentially complete within 2 h. The

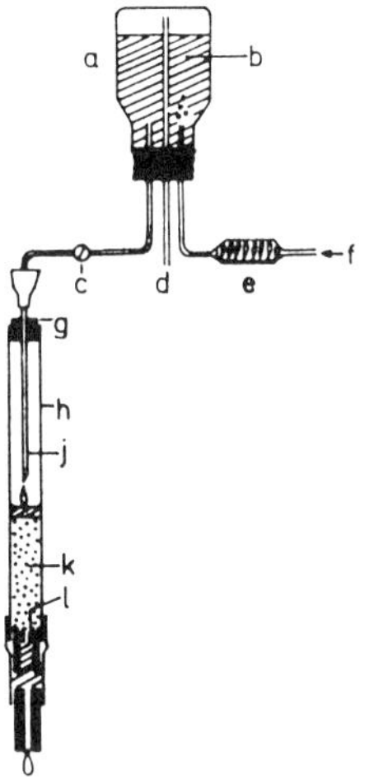

Figure 1. Microenvironment (column system) used for DNA adsorption studies and transformation experiments.
a) reservoir; b) buffer or salt solution or cell-free filtrate of a competent culture (aerated); c) peristaltic high precision pump; d) air outlet; e) cotton filter; f) inlet of compressed air; g) rubber plug; h) glass column; i) syringe; k) sea sand; l) cellulose filter

amount of DNA adsorbed increased with the concentration and valency of cations (Tab. 2). In the presence of 0.2 M $MgCl_2$, adsorption increased also with pH between 5 and 9, with a minimum at pH 7. The association of DNA and sand was relatively stable, since only about 50% of the adsorbed DNA was released during 6 h of elution with the loading buffer including 1 M NaCl. With 0.4 M EDTA solution, 60% of DNA (bound in the presence of 0.2 M $MgCl_2$) were released within 60 min. Detergents did not promote desorption. The results (22) suggest that in the presence of Na^+ DNA adsorbs by means of physical attraction forces (van der Waals) whereas with Mg^{++} DNA adsorbs by bridging interactions between phosphate and silicate anions. Plasmid DNA did also bind to sand.

Table 2: Adsorption of DNA (calf thymus) to sea sand

Condition (2 h at 23°C, pH 7.0)	µg DNA bound per 0.7 g sand
0.0195 M Na^+ (0.1 x SSC)	1.2
0.1 M NaCl	4.8
1.0 M NaCl	7.7
4.0 M NaCl	13.4
0.001 M $MgCl_2$	2.5
0.005 M $MgCl_2$	5.3
0.05 M $MgCl_2$	12.3
0.2 M $MgCl_2$	14.5
0.05 M $CaCl_2$	15.9

Increased resistance of adsorbed DNA to enzymatic degradation:

Previous work has shown that DNA becomes more resistant to DNase I when bound to sand than when free (26). The sand does not inhibit DNase I activity (26). As a sensitive means for monitoring residual biological activity of DNA treated with DNase I we used transformation of *Bacillus subtilis*. Fig. 2 indicates that transforming DNA adsorbed to sand in the presence of 30 mM $MgCl_2$ retained 50% of its activity after a treatment with DNase I (10 ng/ml). Less than 1% of free DNA survived the same treatment. From further inactivation kinetics of DNA adsorbed at 20 to 50 mM $MgCl_2$ it was concluded that two modes of DNA-binding to sand (86.1% quartz) exist which result in different DNase I sensitivities (22). It was also shown that binding and protection of DNA occured on minor mineral fractions of sand, i.e. feldspar and heavy minerals (22). However, these components of sand exhibited types of protection different from quartz in relation to $MgCl_2$ concentrations present during DNA adsorption. Apparently, the various mineral components of sand contribute all to adsorption and protection of DNA but differently with respect to ionic conditions.

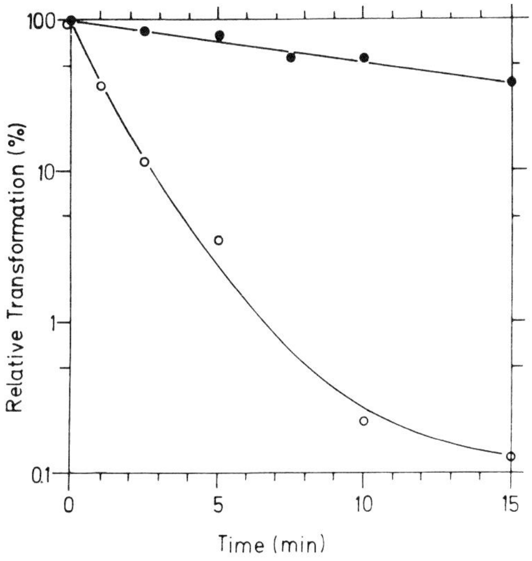

Figure 2. Kinetics of inactivation by DNase I of transforming DNA from *B. subtilis* adsorbed to sand (●) or free in solution (○)

Transformation on sand grains:

Compared to standard transformation of *B. subtilis* in liquid medium (DNA saturation conditions) the overall transformation on sand grains was 25 to 50 fold more efficient (Tab. 3). This remarkable observation gains even more

Table 3: Transformation of *B. subtilis* 1G20 *trpC2* at 23°C and a DNA concentration of 13.3 µg/ml

Condition		Relative transformation efficiency
1.	Standard transformation (DNA plus cells in liquid medium)	1[a)]
2.	Transformation at sand:	
2.1	- DNA adsorbed to sand, then cells added; total cells	25
	- sorbed cells considered only	100
2.2	- Cells adsorbed to sand: then DNA added; total cells	50
	- sorbed cells considered only	3200

a) 0.5×10^{-4} Trp^+ transformants per viable cell impressive significance when only those cells were considered which, after incubation, were recovered from sand by vortexing. Then up to 16% of sorbed cells appeared as transformants (Tab. 3). The experiments were designed and evaluated to give additional information (23). Of the cells applied to the microenvironment about 10% were retained by the DNA-loaded sand and only about 1% by sand without DNA. Furthermore, in the major fraction of cells, which appeared in the effluent from DNA-loaded sand the majority of transformants was recovered. This may indicate that cells take up DNA from sand and

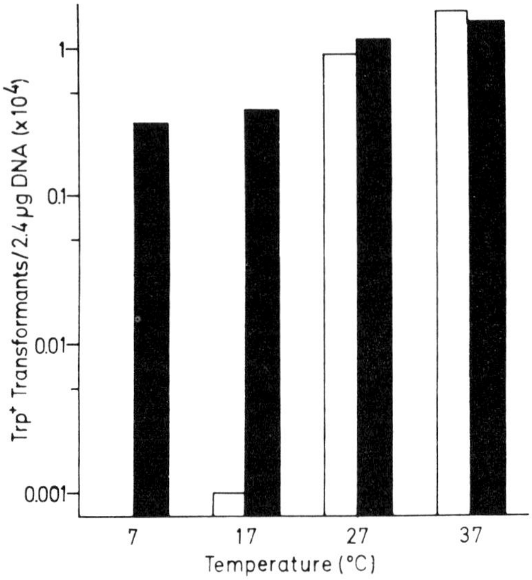

Figure 3. Influence of temperature on transformation of B. subtilis. Cells were incubated at the indicated temperature for 30 min with free DNA (white bars) or with DNA adsorbed to sand grains (black bars)

then detach. The binding of cells to sand observed during these experiments was shown to be a reversible attachment to negatively charged surfaces (23). Scannin electronmicroscopy showed that cells were located on exposed sites of different minerals and were not entrapped in holes or dents (23). Finally, transformation at sand particles was unaffected by DNase I (1 µg/ml), a concentration that essentially abolished transformation in liquid medium. At low temperatures trans formation at sand grains was 100 to 1000 fold more efficient than in liquid culture (Fig. 3). This finding is particularly interesting since it is not likel that in nature the optimum temperature for transformation (37°C) will be regularly found.

Discussion

Mineral surfaces are an important component of bacterial environments, particular for species living in soil and sediments. The association of cells with such surfaces constitutes a normal, if not essential, state in the life cycle of these bacteria. We have studied some aspects of gene transfer by transformation in a microenvironmental model system in which sea sand provided the mineral surfaces. These studies revealed that surfaces of quartz, feldspar and heavy minerals have multiple impacts on gene transfer by transforming DNA:

(i) Stabilization of DNA by adsorption. Earlier observations made with sea sand (26, 27) were extended to the finding that DNA adsorbs to the surface of various minerals under a variety of conditions. Once adsorbed, the DNA becomes much more resistant to DNase I.

(ii) Enhancement of transformation. This may be explained by the transition from three dimensions in liquid culture to quasi two dimensional conditions of DNA-cell interactions at surfaces. Also, the observed increased binding of cells to DNA-loaded sand may contribute to the enhancement.

(iii) Changes of the physiology of DNA binding or DNA uptake by cells. These changes are documented by a substantial increase in the DNase I resistance of transformation and by the efficient transformation even at 7°C. Apparently one or several steps of the transformation process are altered and have become cryoresistant at the mineral surface. Transformation below the optimum temperature is possibly of more ecological relevance than transformation at 37°C.

More experiments are required with other bacteria of soil and sediment including other mineral materials present in natural habitats. Also, detailed studies on plasmid DNA transformation under ecologically relevant conditions are now being initiated.

The necessity to study genetic transformation in natural environments derives from the fact that genetic material may not rely on living donor organisms to be transferred. Microbial cells are known to excrete or release DNA during special stages of their life cycle or after cell death and lysis. This DNA may be available for transformation (28, 29). Hence, also non-conjugative plasmids have the option of being spread to other species in the absence of mobilizing helper plasmids. A further point requesting the consideration of transformation in nature as a powerful gene transfer mechanism is the observation that among Bacillus, Azotobacter, and Cyanobacteria interspecies transformation of selected chromosomal genes has been observed (10, 15, 30).

References

1. Stotzky, G. and Babich, H. (1986). Adv. Appl. Microbiol. 31, 93-138.
2. Hall, B.G. and Zuzel, T. (1980). Proc. Natl. Acad. Sci. USA 77, 3529-3533.
3. Graham, J.B. and Istock, C.A. (1979). Science 204, 637-639.
4. Graham, J.B. and Istock, C.A. (1978). Mol. Gen. Genet. 166, 287-290.
5. McLaren, A.D. and Skujins, J. (1968). In: The Ecology of Soil Bacteria (Eds. Gray, T.R.G. and Parkinson, D.), pp. 1-24, University Press, Liverpool.
6. Fletcher, M. (1980). In: Microbial Adhesion to Surfaces (Eds. Berkeley, R.C.W., Lynch, J.M., Melling, J., Rutter, P.R. and Vincent, B.), pp. 197-210, Ellis Horwood, Chichester.
7. Wimpenny, J.W.T., Lovitt, R.W. and Coombs, J.P. (1983). In: Microbes in Their Natural Environments (Eds. Slater, J.H., Whittenbury, R. and Wimpenny, J.W.T.), pp. 67-117, Cambridge University Press, Cambridge.
8. Grigorieva, G. and Shestakov, S. (1982). FEMS Microbiol. Letters 13, 367-370.
9. Lorenz, M.G. unpublished result
10. Stevens, S.E. and Porter, R.D. (1986). J. Bacteriol. 167, 1074-1076.
11. Trehan, K. and Sinha, U. (1981). J. Gen. Microbiol. 124, 349-352.
12. Yankofsky, S.A., Gurevich, R., Grimland, N. and Stark, A.A. (1983). J. Bacteriol. 153, 652-657.
13. Mulder, J.A. and Venema, G. (1982). J. Bacteriol. 150, 260-268.
14. Juni, E. and Janik, A. (1969). J. Bacteriol. 98, 281-288.
15. Page, W.J. (1985). Can. J. Microbiol. 31, 659-662.
16. Tirgari, S. and Moseley, B.E.B. (1980). J. Gen. Microbiol. 119, 287-296.
17. Carlson, C.A., Pierson, L.S., Rosen, J.J. and Ingraham, J.L. (1983). J. Bacteriol. 153, 93-99.

18. Norgard, M.V. and Imaeda, T. (1978). J. Bacteriol. 133, 1254-1262.
19. Hopwood, D.A. and Wright, H.M. (1972). J. Gen. Microbiol. 71, 383-398.
20. O'Conner, M., Wopat, A. and Hanson, R.S. (1977). J. Gen. Microbiol. 98, 265-272.
21. Bertani, G. and Baresi, L. (1987). J. Bacteriol. 169, 2730-2738.
22. Lorenz, M.G. and Wackernagel, W. (1987). Appl. Environ. Microbiol. (in press)
23. Lorenz, M.G., Aardema, B.W. and Wackernagel, W., J. Gen. Microbiol. (in press)
24. Richards, G.M. (1974). Analyt. Biochem. 57, 369-376.
25. Bron, S. and Venema, G. (1972). Mut. Res. 15, 1-10.
26. Lorenz, M.G., Aardema, B.W. and Krumbein, W.E. (1981). Mar. Biol. 64, 225-230.
27. Aardema, B.W., Lorenz, M.G. and Krumbein, W.E. (1983). Appl. Environ. Microbiol. 46, 417-420.
28. Herdman, M. and Carr, N.G. (1971). J. Gen. Microbiol. 68, XIV.
29. Spizizen, J. (1958). Proc. Natl. Acad. Sci. USA 47, 505- 512.
30. Harford, N. and Mergeay, M. (1973). Mol. Gen. Genet. 120, 151-155.

THE USE OF IS-ELEMENTS FOR THE CHARACTERIZATION OF GRAM-NEGATIVE BACTERIA

Reinhard Simon, Bärbel Klauke and Barbara Hötte
Universität Bielefeld, Lehrstuhl für Genetik,
Postfach 8640, D-4800 Bielefeld 1, FRG

Summary

Positive selection procedures for the isolation of transposable DNA elements from Gram-negative bacteria are described. The use of endogenous insertion sequences for the identification and characterization of wild-type bacterial populations is discussed.

Introduction

Insertion sequences (IS-elements) are defined as mobile DNA segments that carry no detectable phenotype unrelated to the transposition functions [1]. As a consequence, IS elements were usually indirectly discovered by their ability to cause insertional mutations. Most data are available about IS-elements isolated from *E.coli* [2,3]. But there is also an increasing number of reports on the detection of spontaneous mutations in other Gram-negative bacteria that are caused by endogenous IS-elements [4-9].
Gay *et al.* [10] have recently published a method allowing positive selection for the insertion of an IS-element into the vector pUCD800 maintained in *Agrobacterium tumefaciens*. This vector contains the *sacB* gene from *Bacillus subtilis* and prevents cell growth in the presence of sucrose unless this marker is inactivated.
We report here two alternative vector systems that can be used to identify the insertion of IS-elements by drug resistance selection.

Risk Assessment for Deliberate Releases
Edited by W. Klingmüller
© Springer-Verlag Berlin Heidelberg 1988

Results and Discussion

Vector plasmid, mobilization and principle of selection

We have used pSUP104 [11] as the basic replicon for the construction of the IS-isolation vectors.

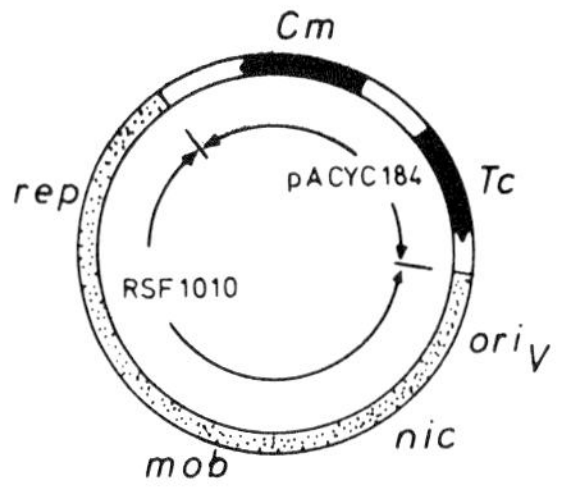

pSUP104 (9,5 kb)

The vector consists of parts of the RSF1010 plasmid including its broad-host-range replication- (*rep*, *ori*$_V$) and mobilization (*mob*, *nic*) functions fused to the drug resistance markers of the *E.coli* vector pACYC184 (Cm: chloramphenicol, Tc: tetracyclin).

Since pSUP104 is not self transmissible we have used a special *E.coli* donor strain (mobilizing strain) [12] that carries a derivative of the broad host range resistance plasmid RP4 integrated into the chromosome. The integrated RP4 plasmid is devoid of its natural resistance genes but still expresses the transfer functions necessary to mobilize the vector pSUP104.

E.coli donor non-*E.coli* recipient

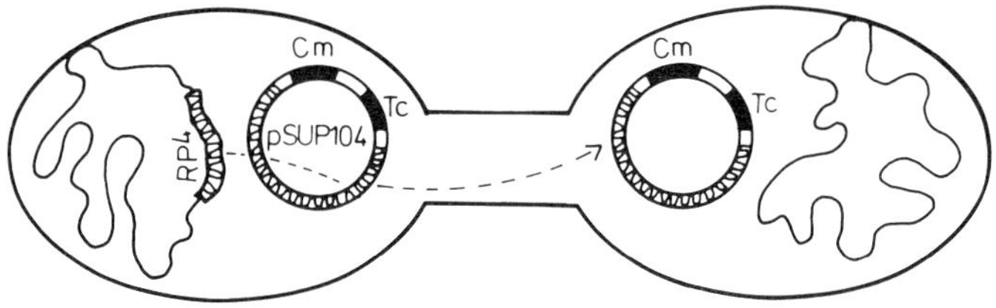

Recipients tested:

Rhizobium meliloti, Rhizobium leguminosarum, Rhizobium trifolii, Agrobacterium tumefaciens, Pseudomonas aeruginosa, Pseudomonas stuzeri, Rhodobacter capsulatus, Xanthomonas campestris

The vector is stably maintained in the recipients mentioned, the mobilization frequency of pSUP104 into these non-*E.coli* recipients can reach up to 100% (11).

The principle of the selection procedure can be summarized as follows.

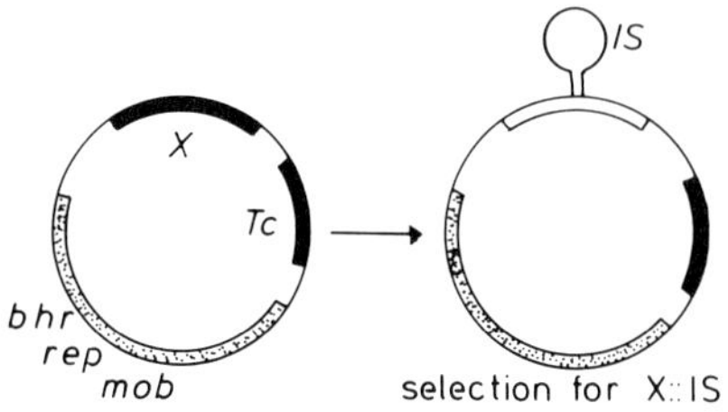

An indicator gene (designated X) is inserted into pSUP104. Then, the vector is mobilized from *E.coli* into the recipient of interest to allow IS-element insertions to occur. Gene X encodes for a dominant phenotypical trait, it confers to the recipient sensitivity to a particular antibiotic for which there is a chromosomally located, recessive resistance determinant. Consequently, the insertion of an IS-element into gene X (X::IS) can be positively selected by the appropriate antibiotic.

The indicator genes

We have used two different genes to carry out the experiments outlined above. Both indicator genes originate from *E.coli* and were taken from cloning vectors originally designed for *in-vitro* insertion of DNA-restriction fragments.

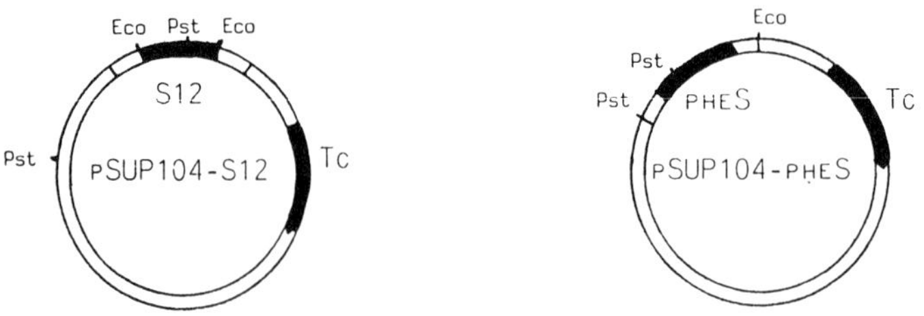

Vector pSUP104-pheS contains the *E.coli pheS* gene that encodes a subunit of the phenylalanine-t-RNA-synthetase [13]. The *E.coli pheS*-mutant RR28 is resistant to the drug parafluorophenylalanine (pfp) [13], but the vector borne wildtype *pheS* gene is dominant so that RR28 harbouring pSUP104-pheS is sensitive to this antibiotic. In order to use pSUP104-pheS as an IS-selection vector it is first mobilized into the non-*E.coli* recipient and, after beeing maintained there for some time, transferred back to *E.coli* RR28 by the help of an Tc^S derivative of RP4. In RR28 the selection for both, Tc^r and pfp^r, allows the detection of vectors that carry an IS-element inserted in the pheS-gene.
The second system is much simpler since the selection for the marker-gene inactivation can be performed directly in the non-*E.coli* strain of interest. Vector pSUP104-S12 carries the *E.coli rpsL* (or S12) gene encoding a protein of the small subunit of the ribosome [14]. A chromosomal mutation resulting in resistance to streptomycin (Sm^r) is dominated by the vector borne wildtype S12 gene so that a Sm^r strain harbouring pSUP104-S12 is Sm^S unless the indicator gene is inactivated, i.e. the possible insertion of an IS-element into the cloned S12 gene may be selected by Sm resistance.

Isolation of IS-elements

We have successfully tested the above described vectors in a number of different *Rhizobium* strains and some other Gram-negative bacterial species. After appropriate selection, inactivation of the indicator genes was found with frequencies ranging from 10^{-5} to 10^{-8}. Since inactivation of the vector borne *pheS* or S12 gene could have occurred not only by IS-insertions but also by deletions or point mutations, the lengths of the resulting vectors were tested by in-gel-lysis procedures and electrophoresis to screen for enlarged molecules, i.e. vectors that contained DNA-insertions of detectable sizes. The results obtained varied considerably, depending on the strain used in the experiment. In some strains almost all vectors contained an insertion, whereas in other strains the indicator genes had been inactivated predominantly by deletions or point mutations.

So far we have isolated and characterized by restriction analyses more than twenty different IS-elements from various Gram-negative soil bacteria. The sizes of these transposable elements varied in the range from about 1 kb to more than 8 kb.

Hybridization experiments

To verify the origin of the IS-elements isolated as described above, they were used as probes and hybridized to total DNA of the parental strains and to a number of more or less related other bacterial strains.

One example for such experiments is shown.

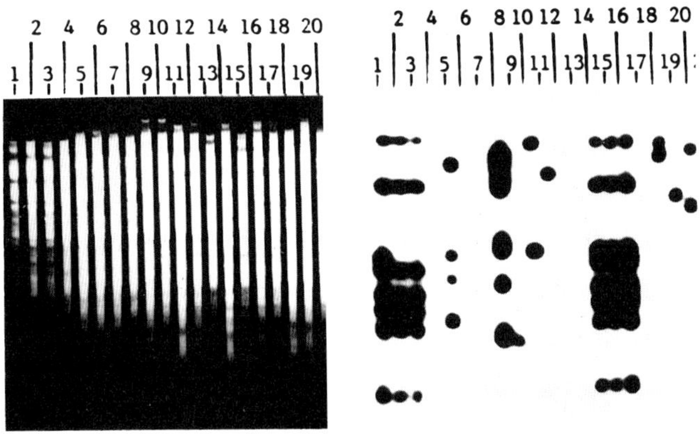

Rhizobium meliloti: 1:2011-Sm (Bielefeld), 2:2011 (Denarie), 3:2011 (Signer), 4:20-1, 5:102F34, 6:102F51, 7:AK632, 8:NGR185, 9:NGR43-4, 10:USDA1024, 11:USDA1029, 12:USDA1082, 13:USDA1010, 14:MM2B, 15:MM6B, 16:MM7C, 17:102F65, 18:MVII, 19:220-12, 20:220-13

The 1.4 kb IS-element ISRm2011-2 used as probe originated from *Rhizobium meliloti* 2011-Sm (lane 1). The other lanes contain total DNA from various other *R.meliloti* strains. The hybridization pattern obtained can be used to classify the bacterial strains. The first 3 strains are identical (except for one additional copy of the IS-element in lane 1). The strains MM2B, MM6B and MM7C, the origin of which was not known to us before, are identical to 2011-Sm. This conclusion was confirmed by similar experiments with other IS-element probes. The other *R.meliloti* strains show significant variations in their hybridization patterns. So far, in only few cases IS-elements isolated from *R.meliloti* were found to hybridize to total DNA of other Rhizobia like *R.leguminosarum* or *R.trifolii*.

Conclusions

We have shown that IS-elements can be isolated very easily by the positive selection vectors described. Hybridization experiments result in clear-cut patterns allowing not only to reveal the number and distribution of the IS-elements but also to analyze quickly and unambiguously the identity of a bacterial strain. We would suggest therefore that this could be an additional method for the detailed characterization of wild-type populations of bacteria, e.g. before or after release of genetically manipulated bacteria into a particular environment.

References

[1] Bukhari, A.I., Shapiro, J.A., and Adhya S.L. (ed) (1977). DNA insertion elements, plasmids, and episomes. Cold Spring Harbor Laboratory, Cold Spring Harbor, N.Y.

[2] Shapiro, J.A. (ed) (1983). Mobile genetic elements. Academic Press Inc., N.Y.

[3] Starlinger, P, and Saedler, H. (1976). Curr. Top. Microbiol. Immunol. 75: 111-153

[4] Priefer, U.B., Burkhardt, H.J., Klipp, W., and Pühler, A. (1981). Cold Spring Harbor Symposia on Quantitative Biology XLV, pp. 87-91

[5] Machida, Y., Sakurai, M., Kiyokawa, S., Ubasawa, A., Suzuku, Y., and Ikeda, J.E. (1984). Proc. Natl. Acad. Sci. USA 81:7495-7499

[6] Comai, L., and Kosuge, T. (1983). J. Bacteriol. 154:1162-1167

[7] Scordilis, G.E., Ree, H., and Lessie, T.G. (1987). J. Bacteriol. 169:8-13

[8] Ruvkun, G.B., Long, S.R., Meade, H.M., VandenBos, R.C., and Ausubel, F.M. (1982). J. Mol. Appl. Genet. **1**:405-418

[9] Dusha, I., Kovalenko, S., Banfalvi, Z., and Kondorosi, A. (1987). J. Bacteriol. **169**:1403-1409

[10] Gay, P., LeCoq, D., Steinmetz, M., and Kado, C.I. (1985) J. Bacteriol. **164**:918-921

[11] Priefer, U.B., Simon, R., and Pühler, A. (1985). J. Bacteriol. **163**:324-330

[12] Simon, R., Priefer, U.B., and Pühler, A. (1982). Biotechnology **1**:784-791

[13] Hennecke, H., Günther, J., and Binder, F. (1982). Gene **19**:231-234

[14] Dean, D. (1981). Gene **15**:99-102

BIOLOGICAL CONTAINMENT OF BACTERIA AND PLASMIDS TO BE RELEASED IN THE ENVIRONMENT

S. Molin, P. Klemm, L.K. Poulsen*)
H. Biehl, K. Gerdes, P. Andersson*)
Department of Microbiology, The Technical University of Denmark, building 221, DK-2800 Lyngby

*) The Genetic Engineering group, The Technical University of Denmark, building 227, DK-2800 Lyngby

Summary: The design of a biological containment system to be employed in a broad spectrum of bacteria useful in connection with release to the external evironment is reported. The key element is a gene, hok, encoding a small polypeptide of 52 amino acids which, when expressed, is lethal to several different bacterial species. Through construction of various combinations of regulatable promoters and this toxin gene we have achieved containment, since cells accidentally escaping to the outside environment, or any other new combination of a bacterium and the toxin carrying plasmid, will be killed. A specific application of the containment system with respect to deliberate release is based on a fusion between an invertible promoter (flip-flop sequence) and the toxin gene. This results in a stochastic induction of the killing function which eventually will lead to the non-conditional elimination of the organism as a consequence of the competition with related bacteria in the environment. Key words: Containment, invertible promoter, deliberate release.

Introduction

Since the beginning of the genetic engineering era biological containment of recombinant DNA (rDNA) bacteria has been considered important and normally involves the use of specifically debilitated organisms and non-conjugative, non-mobilizable plasmids (1,2).

This approach to biological containment has certain shortcomings. Firstly, the debilitating mutations may negatively affect the growth properties of the organisms. Secondly, an elaborate mutant selection may be necessary to obtain an organism which is properly contained, yet has the desired growth properties. Thirdly, containment of intentionally released rDNA organisms is hardly feasible by introduction of debilitating mutations since the establishment and maintenance of released rDNA organisms require that the rDNA organisms are able to compete favorably with wild-type organisms. Finally, nonmobilizable plasmids may be transferred to wild-type bacteria in the environment by bacteriophage mediated transduction, recombination with a superinfecting conjugative plasmid, or passive uptake (transformation) of plasmid DNA released from rDNA organisms (3,4).

Risk Assessment for Deliberate Releases
Edited by W. Klingmüller
© Springer-Verlag Berlin Heidelberg 1988

We have developed a novel strategy for biological containment of bacteria as well as plasmids. A DNA "cassette" is incorporated into the organisms which consists of a gene encoding a cell killing function and such regulatory sequences that a conditional suicide system is created. This strategy of containment is applicable to organisms growing in fermenters, organisms to be released, and to recombinant DNA vectors.

Materials and methods

Bacterial strains used for cloning were *E. coli* MC1000 (5) (*araD139*) (*araABOIC leu*)7697 (*lacIPOZY*)X74 *galU galK rpsL thi*) and *B. subtilis* BD170 (6) (*trpC2 thr-5*).

Media employed were LB medium (7), A+B minimal medium (8), and glycerol minimal medium supplemented with 10ug/ml threonine, isoleucine, proline, leucine, and 0.2% glycerol. LB plates were made by addition to LB medium of agar to 1%.

DNA manipulations and *E. coli* transformations were carried out as described (9). Transformation of *B. subtilis* BD170 was done as described (10). DNA sequence determination was performed using the dideoxyribonucleotide chain termination method on supercoiled plasmid DNA (11). Primers for sequencing and oligonucleotides used for cloning were synthesized by the phosphoramidite method using the Cyclon DNA® synthesizer.

Results

The bacterial cell killing function

The cell killing function to be incorporated as the effector part of an active biological containment system must fulfill certain requirements. The gene encoding the cell killing function should be contained within a small DNA fragment, so that the insertion of the fragment into a production plasmid does not adversely affect the copy number of the plasmid by increasing its size. A second criterion is that the cell killing function should be active in a broad range of bacteria if the system is to be used for containment of a production plasmid and/or in various bacterial species.

We have chosen to investigate the applicability of a cell killing function derived from the *E. coli* plasmid R1. The R1 plasmid contains a region, *parB*, which when inserted into a plasmid ensures the stable maintenance of the plasmid in a bacterial population as cells, from which the *parB* containing plasmid is lost, are killed by derepression of a cell killing function, R1 *hok* (12). The R1 *hok* gene product is a small protein of 52 amino acids which presumably acts at the cell membrane since the immediate effect of *hok* expression is a collapse of of the transmembrane potential (13). As shown in Fig. 1 induction of the *hok* gene under control of a regulatable promoter (λP_R) leads to a rapid killing of the cells.

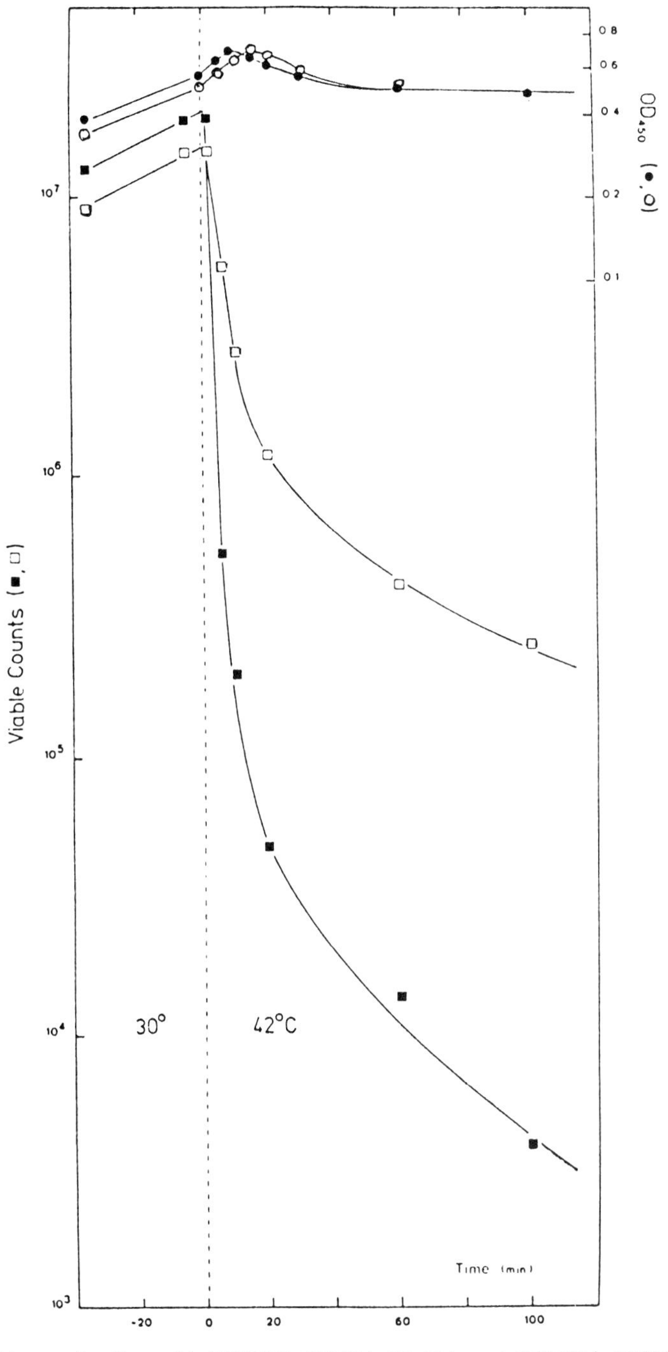

Figure 1. E. coli MC1000(pKG341) (O, □) and MC1000(pKG345) (●, ■) were grown in LB medium with ampicillin (50 ug per ml) at 30°C to OD_{450}=approx. 0.5 at which time the temperature was increase to 42°C. OD_{450} and viable counts were followed. The latter was determined by plating samples on LB plates containing ampicillin with incubation at 30°C. The plasmids pKG341 and pKG345 have λP_R inserted at different positions upstream of hok.

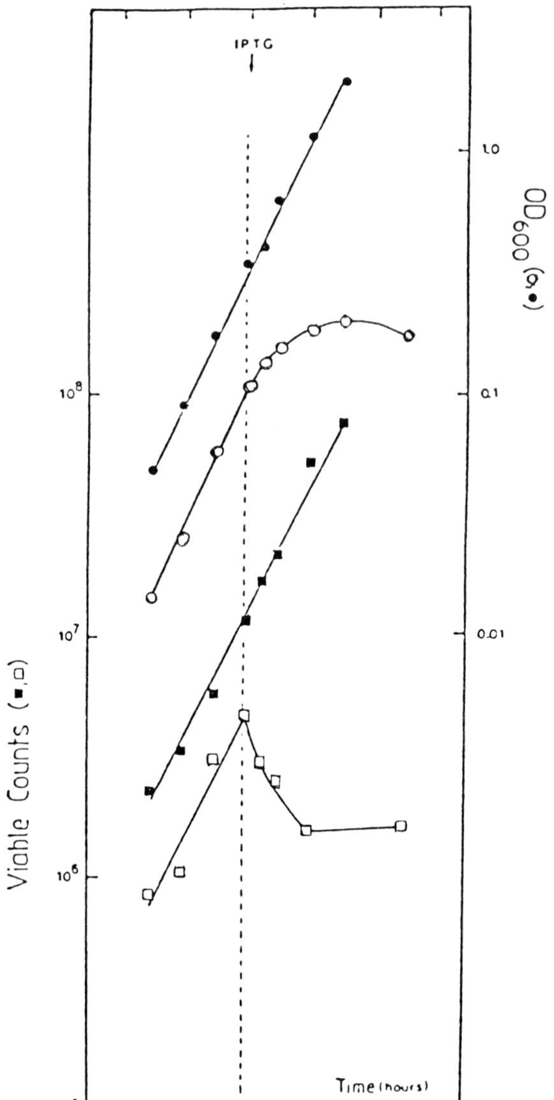

Figure 2. B. subtilis BD170(pLK26) (○,□) and BD170(pSI-1) (●,■) were grown at 37°C in LB medium. At the time indicated, IPTG was added to a final concentration of 2mM to induce expression from the P_{spac-I} promoter of pSI-1. OD_{600} and viable counts was followed, the latter by plating samples on LB plates with chloramphenicol.

Since the target for the Hok protein seems to be located in the cell membrane, it was contemplated that the target may be of a universal nature and, consequently, that the hok killing function might be active in a broad range of bacteria. We therefore analyzed the effect of R1 hok expression in the Gram positive B. subtilis.

The plasmid used in these experiments was pLK26. This plasmid was constructed from pSI-1 (14) which is an E. coli-B.subtilis shuttle vector containing a B. subtilis bacteriophage promoter with the E. coli lac operator (spac-I promoter) as well as the E. coli lacI gene specifying the lac repressor. Genes inserted into pSI-1 can be expresssed in B. subtilis following isopropyl-β-D-thiogalactosidase (IPTG) induction of P_{spac-I}. Part of the R1 hok coding sequence together with a synthetic oligonucleotide encoding the N-terminus of Hok was inserted into pSI-1. When B. subtilis BD170 harbouring pLK26 was induced with IPTG (Fig 2.), the cells stopped growing and viable counts immediately decreased by a factor of 4-5. By phase contrast microscopy, cells with abnormal morphology could be observed from one hour following IPTG induction. These "ghost"-like cells resemble the ghost cells described earlier for Hok killed E. coli (12). Cells surviving IPTG induced hok expression were found to contain an intact and functional hok gene by a second challenge with IPTG as above. We therefore conclude that the R1 hok cell killing function is active also in B. subtilis albeit at a low efficiency which, however, may be increased by substituting the promoter or the Shine-Dalgarno sequence.

The data on expression of hok in B. subtilis, together with evidence presented elsewhere (15) indicating that the hok protein is toxic in a broad range of Gram negative bacteria, suggest that the hok gene may be an excellent candidate for an effector gene to be incorporated into a containment cassette.

A model system for biological containment of bacteria to be released.

We propose a containment principle which relies on imposing a growth disadvantage of the released population as such, rather than on debilitation of the individual cell. It is suggested that this can be achieved by inserting a suicide function which is activated at random thereby resulting in killing of a fraction of the released cells per time unit.

A plasmid, pPKL100, was constructed in which the R1 hok gene is activated at random by the E. coli fimA promoter. This promoter is located on a 300 bp invertible DNA segment that specifies a periodic expression of type 1 fimbriae in E. coli (16). The inversion is controlled in trans by two regulatory genes, fimB and fimE (17), which act antagonistically. The FimB protein mediates an "on" configuration of the switch in which fimA is transcribed, and the latter an "off" configuration. Plasmid pPKL100 contains one copy of each of the fimB and fimE genes (Fig. 3).

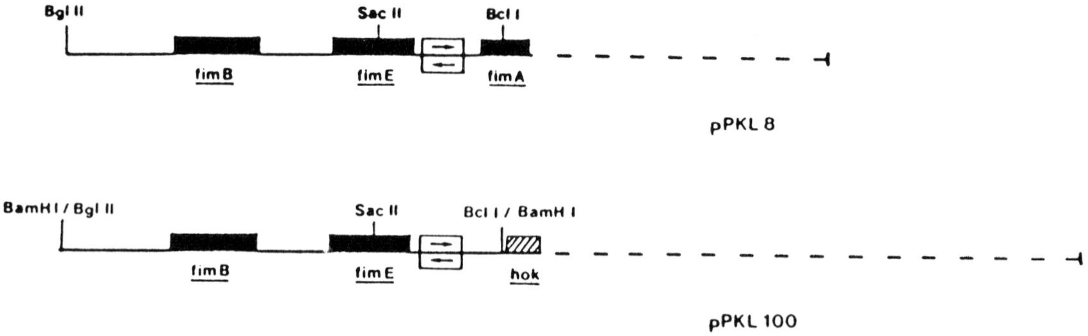

Figure 3. The pBR322 derivative pPKL8 carries part of the fim gene cluster from E. coli: the invertible segment containing the fimA promoter a truncated fimA gene, and two genes, fimB and fimE, whose gene products are involved in the regulation of the frequency of inversion of the fimA promoter.

The 3.3 kbp BglII-BclI fragment from pPKL8 (17) which includes the fimA promoter segment and one copy of each of the fimB and fimE genes was inserted upstream for the hok Shine-Dalgarno sequence to generate pPKL100 in which hok expression is regulated by the fimA promoter.

Clones of E. coli MC1000 were made, which in addition to pPKL100 (a pBR322 replicon), harboured either one of three compatible pACYC184-derived plasmids, pPKL104 (fimB$^+$), pPKL105 (fimB$^+$, fimE$^+$), or pPKL106 (fimE$^+$). Since the ratio of the copy numbers for pBR322 derivatives and pACYC184 derivatives is approximately four to one, the relative gene dosage of fimB and/or fimE is increased by 25% in the different strains. Since the ratio between fimB and fimE is not changed in MC1000(pPKL100, pPKL105), this strain was considered a suitable control.

Overnight cultures of E. coli MC1000 harbouring either pPKL100 alone, or pPKL100 together with pPKL104, pPKL105, and pPLK106 contained dead ghost cells characteristic for R1 Hok killing at levels of 1-2% (pPKL100 and pPKL100 + pPKL105), 10% (pPKL100 + pPKL104) and 0.1% (pPKL100 + pPKL106), i.e. ghost formation is a function of the switch frequency of the fimA promoter.

The population doubling time for cells carrying the fusion between the fimA promoter and hok was determined in glycerol minimal medium. E. coli MC1000(pPKL100) was compared to control strains harbouring either the hok gene without a promoter(pPR341) or the fimA promoter (pPKL8), Table 1. The population doubling time for MC1000(pPKL100 was increased by approx. 10% relative to the control strains MC1000(pPKL8) and MC1000(pPR341). The presence of pPKL100 thus leads to a significant growth disadvantage for the population as such, while the individual living cell is unaffected by the presence of a non-expressed hok gene.

Table 1. Doubling time, as measured by increase in optical density of MC1000 cells harbouring various plasmids

Plasmids	fimB:fimE ratio	Doubling time (min.) in glycerol minimal medium
pPKL100	20:20	95
pPKL8	20:20	85
pPR341	-	86
pPKL100 + pPKL104	25:20	130
pPKL100 + pPKL105	20:20	118
pPKL100 + pPKL106	20:25	96

The presence of a 25% excess of fimB in MC1000(pPKL1000, pPKL104) leads to a further increase in the population doubling time relative to MC1000(pPKL100, pPKL105) which contains equal dosages of fimB and fimE (Table 1). In contrast, a decrease in population doubling time is seen, if an excess fimE is present in trans, cfr. MC1000(pPKL100, pPKL106) compared to MC1000(pPKL100, pPKL105). It is therefore possible by varying the ratio of fimB and fimE expression to design different containment cassettes based on fimA-activated hok expression which at a given generation time will result in killing of a predetermined fraction of the cells per time unit.

Discussion

Although one may envisage a number of combinations of regulatory promoters and toxic or lethal genes in bacteria, the system described here seems to have a couple of advantages which may be difficult to obtain with many other systems. One is the size of the toxin gene, hok, which, because it only constitutes approx. 200 bp, may be inserted into hybrid plasmid without serious effects on copy number and gene expression. Also the efficient killing activity of the Hok protein is favourable compared with growth inhibition when being part of a containment system. Finally, the apparent broad spectred activity in many different bacteria makes hok a possible tool for containment of new bacteria used in industry or agriculture, and also an improved plasmid containment gene.

Our aim in the use of fusions between the flip-flop fimA promoter and the hok gene is to design a containment principle to be combined with organisms that are to be placed deliberately in the external environment. Our preliminary results indicate that this type of stochastic killing of individuals in a population of cells introduces an element of "aging" which eventually will lead to the elimination of these cells. The major advantage of this regulatory system in connection with its use in the outside environment is that there is no need for any external agents (chemicals, temperature etc.) to activate the killing function - it occurs spontaneously with a low but significant frequency.

Following up on the growth rate determinations presented here we also have preliminary data showing that in a mixed population of *E. coli* cells with and without the flip-flop containment, competition favors the wild-type without the system, showing that although a complete elimination is not achieved by this type of containment system, the cells will nevertheless in the long run be competed out by their wild-type relatives.

Finally, it should be added, that in suspensions of non-growing cells left for many days without addition of new energy ressources a continuous reduction of the cell number is observed if a flip-flop-*hok* system is present. In one preliminary experiment viable counts dropped by a factor of 10^6 over a period of 9 days under such stationary conditions (the control strain was almost fully viable).

It thus seems that the inversion frequency of the *fimA* promoter is time dependent rather than growth rate dependent, which probably means that the containment system is operating under all sorts of environmental conditions.

Acknowledgements

We are grateful for the financial support from the Danish Ministry of the Environment.

References

1. Federal register, Department of Health and Human Services. Part III: Guidelines for research involving recombinant DNA molecules. National Institutes of Health (1986), 16957-16985.

2. Curtiss, R., III, Inoue, M., Pereira, D., Hsu, J.C., Alexander, L., and Rock, L. (1977). In Scott, W.A. and Werner, R. (eds.) Molecular Cloning of recombinant DNA, Academic Press, New York, p. 248-261.

3. Slater, J.H. (1985). In Halvorson, H.O., Pramer, D. and Rogul, M. (eds.) Engineered Organisms in the environment: Scientific issues, ASM, Wash.DC, p. 89-98.

4. Shapiro, J.A. (1985). In Halvorson, H.O., Pramer, D. and Rogul, M. (eds.) Engineered organisms in the environment: Scientific issues, ASM, Washington, DC, p. 63-69.

5. Casadaban, M. and Cohen, S.N. (1980). J. Mol. Biol. 138, 179-207.

6. Dubnau, D. and Cirigliano, C. (1974). J. Bacteriol. 117, 488-493.

7. Bertani, G. (1951). J. Bacteriol. 62, 293-300.

8. Clark, J.D. and Maaloe, O. (1967). J. Mol. Biol. 23, 99-112.

9. Maniatis, T., Fritsch, E.F. and Sambrook, J. (1982). Molecular cloning, a laboratory manual. Cold Spring Harbor Laboratory.

10. Sadaie, Y. and Kada, T. (1983). J. Bacteriol. 153, 813-821.

11. Promega Notes (1986) no. 5, p. 1-4. Promega Biotec, Madison, USA.

12. Gerdes, K., Rasmussen, P.B. and Molin, S. (1986). Proc. Natl. Acad. Sci. USA 83, 3116-3120.

13. Gerdes, K., Bech, F.W., Jorgensen, S.T., Lobner-Olesen, A., Rasmussen, P.B., Atlung, T., Karstrom, O., Molin, S. and von Meyenburg, K. (1986). EMBO J. 5, 2023-2029.

14. Yansura, D.G. and Henner, D.J. (1984). Proc. Natl. Acad. Sci. USA 81, 439-443.

15. Molin, S., Klemm, P., Poulsen, L.K., Biehl, H., Gerdes, K. and Andersson, P. (1987). Biotechnology, in press (December issue).

16. Abraham, J.M., Freitag, C.S., Clements, J.R. and Eisenstein, B.I. (1985). Proc. Natl. Acad. Sci. USA 82, 5724-5727.

17. Klemm, P. (1986). EMBO J. 6, 1389-1393.

DETECTION OF CONTAINMENT BREACH IN BIOPROCESS PLANT USING AEOROBIOLOGICAL MONITORS

I W Stewart and G Leaver
Biotechnology and Separations Division
Warren Spring Laboratory, Stevenage, Hertfordshire, UK

Summary: This paper describes the progress made on the UK part of this joint project on the assessment of risks in biotechnological processing (BAP 110 UK). One objective of the work is to assess and compare the suitability of air sampling monitors for detecting breaches of containment during bio-processing. Three viable samplers and two clean room monitors were used to sample an aerosol of a spore suspension generated in a containment cabinet. Differences in collection efficiency are explained in terms of droplet size effects. Future and related work is briefly discussed.
Keywords: aerobiological monitoring, aerosols, bioprocess plant, safety, containment, droplet size measurement.

Introduction

The rapid development of biotechnology by the introduction of new techniques has raised issues such as the perceived risk of operator exposure to the microorganisms and their products. Recent international guidelines (1) have concentrated on the risks associated with genetically engineered organisms both categories have been generally established for organisms, the translation of the guidelines into process plant details still requires individual judgement and verification by the regulatory authorities and process operators. One recommendation by OECD for good occupational hygiene is to test for the presence of viable process organisms outside the primary physical containment (process equipment). This practice would ensure that relevant data are built up on exposure levels and reliability of process plant.

The exposure limits and emissions from products of biotechnology are equally important particularly with biologically active compounds. The debate on risks associated with such products is not so advanced as the risk from

Risk Assessment for Deliberate Releases
Edited by W. Klingmüller
© Springer-Verlag Berlin Heidelberg 1988

microorganisms. Nevertheless, it is desirable that more is known about product exposure and downstream process equipment containment so that expensive secondary containment precautions such as glove boxes and operator "moon suits" can be avoided, or reduced (2).

Clearly, primary containment by the process equipment itself should be a principal requirement for protection from harmful exposure to biological materials. While it is not desirable to overdesign equipment to offer complete containment in all cases, it is worthwhile assessing the potential emissions from process equipment to balance this against the exposure limits acceptable for the category of biological material involved. Containment of processing equipment is often determined by methods such as leak testing with helium. However, methods based on collection and analysis of biological materials can provide a more realistic picture of the emissions from equipment during normal operation. Techniques based on particle counting also have an important function since rapid or instantaneous results can be obtained and corrective action can be implemented.

The general objectives of the joint project are as follows:

1. Assess and compare existing methods for the detection of containment breach,

2. Assess containment of process components and unit operations,

3. Assess failure modes in order to improve design,

4. Assess the relevance of existing reliability and risk data bases to bioprocessing,

5. Make information available so that equipment selection can be based on reliability, containment, asepsis and safety.

The work described in this paper was designed to meet the first general objective listed above. There is a range of available aerobiological and other monitors which could be used for process equipment containment evaluation. However, their efficiencies vary widely and little independent scientific comparison has been carried out. This present work has been undertaken to assess the suitability of both aerobiological and particle counting techniques for containment evaluation of process equipment during operation.

Experimental

A containment cabinet constructed to a Warren Spring Laboratory design specification was supplied by Bassaire Ltd (Southampton, UK). The containment cabinet was equipped with High Efficiency Particulate Air (HEPA) filters at each end and provided a constant flow of sterile air within the chamber. An aerosol injection port and a number of sampling ports were incorporated such that simultaneous sampling with several detectors could take place. In this study, three viable microorganism samplers and two clean air monitors have been investigated. The main features of these samplers are summarised in Table 1. Initially the cabinet was validated by injecting sterile air and sampling with all five samplers. After modifications were made to the equipment and sampling probes the comparative studies were carried out.

TABLE 1. - Main Features of Air Samplers

Name (Abbreviation)	Type	Principle	Air Flow Rate litre/min	Units
Casella Slit Sampler (CSS)	Viable	Microorganisms directed through slit above rotating collection plate	30	CFU/litre
Millipore Membrane Filter (0.45 or 0.8 μm) (MMF)	Viable	Air filtered onto membrane	13.4	CFU/litre
Anderson Microbial Sampler (AMS)	Viable	Series of perforated plates mounted above collection plates	28.3	CFU/litre
TSI Laser Particle Counter (LPC)	Clean Air	Particles flow through and interrupt a laser beam	2.83	#/litre
California Measurements Quartz Crystal Monitor (QCM)	Clean Air	Series of orifices mounted above piezoelectric crystals	0.25	μg/litre

CFU = Colony forming unit

The monitors were challenged with spores of Bacillus subtilis var niger (NCIB 8058) produced in a 2 litre batch fermenter in a lean nutrient broth based media (3). After about 48 hours growth, when the nutrients were exhausted, sporulation was initiated by heat treatment of the broth to 70°C for 30 minutes. The spores (at known concentration) were aerosolised in phosphate buffer solution using a purpose built glass atomizer. The samplers were operated for different lengths of time according to the manufacturers instructions and for the viable monitors the number of colonies formed on a solid substrate (Tryptone Soya Agar) after incubation were counted manually. The results were normalised by dividing the number of colony forming units (CFU) by the amount of air sampled. The droplet size distribution of the challenge aerosol was varied by altering the atomiser operating conditions. Comparative measurements of the droplet size distributions were made with a six stage Andersen Microbial Sampler and a ten stage California Measurements Quartz Crystal Monitor. The LPC continuously monitored the aerosol concentration and the remaining four monitors were operated when the maximum aerosol concentration was detected. Typically three to thirteen readings were taken for each detector with individual sampling times ranging from 4 to 8 minutes.

Results

Containment Cabinet Validation

Only the AMS and the LPC detected airborne particles. The LPC detected one particle of the size range 0.5-5 micron per 3 litres of air sampled and one particle above 5 micron in 20 litres of air. The AMS detected 1 CFU in 300 litres of air sampled.

Comparison of Monitors

Some of the results of aerobiological sampling produced to date are listed in Table 2 and shown in Figs 1 (droplet size effect), 2 and 3 (reproducibility). For the viable samplers the results are expressed as percentage recovery which is defined as the airborne spore concentration divided by the normalised colony count and expressed as a percentage.

First consider Fig. 1. Two features are worthy of note. For the viable samplers with a mean droplet size of 3.25 micron the maximum percentage recovery (with the AMS) is around 12%. With a mean droplet size of 1.22 micron the maximum percentage recovery is around 55% again with the AMS. In both sets of results the CSS gave the second highest percentage recoveries followed by the MMF.

Duplicate tests (Fig. 2) performed on consecutive days with mean droplet sizes of approximately 3.5 micron yielded droplet counts (#/litre) within 10% and airborne concentrations (micrograms/litre) also within 10%. As can be seen from Fig. 2 the viable samplers showed differences in percentage recoveries of between 50 and 100%. Also, as can be seen from Table 2, the standard deviation of results with the tests were higher for the larger mean droplet size. Similar tests with a mean droplet size around 1.2 micron (Fig. 3) gave reproducibilities within 25%. As with Fig. 1, in Figs 2 and 3 the viable samplers could be ranked with the AMS giving the highest percentage recovery and the MMF the lowest.

Figures 3 to 8 show the droplet size distributions obtained with the AMS. For the duplicate sets, Figs 5 and 6, and Figs 7 and 8 the distributions are clearly similar. The QCM also gave data on the droplet size distribution. In all cases the mean droplet size given by the QCM was an order of magnitude lower than that given by the AMS. This result suggested that the QCM detected more particles in the sub-micron range.

TABLE 2. - Summary of Results

Date (D_{3v} (AMS))	Sampler	No of Data Points	Mean	Standard Deviation	Theoretical Max Spore Concentration	% Recovery
6.4.87	AMS	6	0.68 CFU/litre	0.04	6.00	11.1
(3.25 µm)	CSS	12	0.38	0.07		6.3
	MMF.8	3	0.18	0.04		3.0
	MMF.45	3	0.13	0.02		2.1
	QCM	7	0.07 µg/litre	0.02	-	-
	LPC	13	64100 #/litre	4600	-	-
8.4.87	AMS	6	1.08 CFU/litre	0.56	6.00	18.0
(3.57 µm)	CSS	12	0.98	0.80		16.3
	MMF.8	4	0.17	0.12		2.8
	MMF.45	4	0.18	0.13		3.0
	QCM	6	0.06 µg/litre	0.01	-	-
	LPC	13	66100 #/litre	3900	-	-
9.4.87	AMS	6	0.76 CFU/litre	0.18	6.00	12.7
(3.49 µm)	CSS	12	0.55	0.19		9.2
	MMF.8	4	0.08	0.05		1.3
	MMF.45	4	0.08	0.01		1.3
	QCM	6	0.07 µg/litre	0.02	-	-
	LPC	11	59500 #/litre	4100	-	-
22.4.87	AMS	6	1.30 CFU/litre	0.11	6.00	54.1
(1.22 µm)	CSS	12	0.95	0.07		39.6
	MMF.45	6	0.54	0.05		22.5
	QCM	6	0.01 µg/litre	0.00	-	-
	LPC	11	28300 #/litre	1500	-	-
23.4.87	AMS	6	1.06 CFU/litre	0.18	6.00	44.2
(1.18 µm)	CSS	12	0.72	0.15		30.2
	MMF.45	6	0.54	0.07		22.3
	QCM	6	0.01 µg/litre	0.00	-	-
	LPC	10	28700 #/litre	1600	-	-

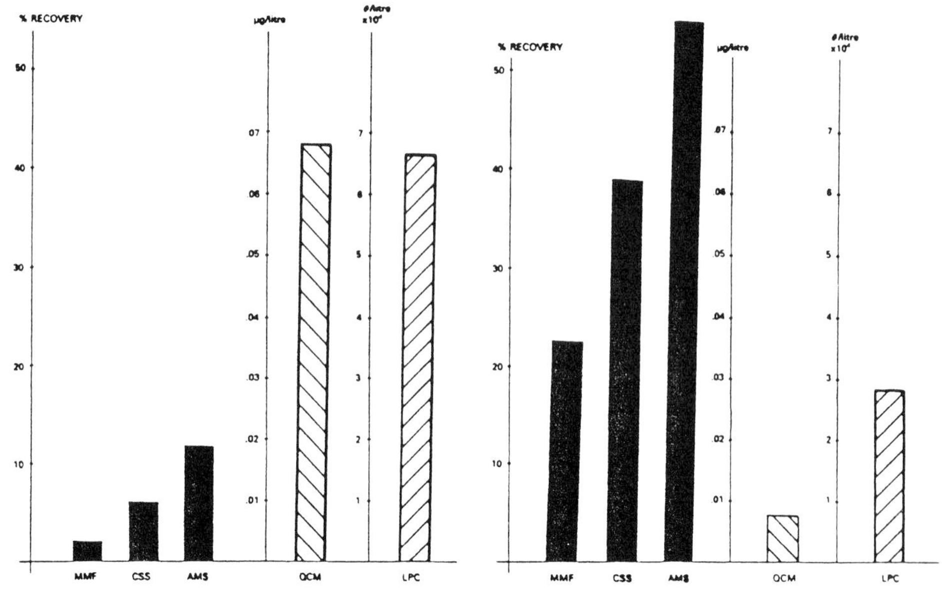

06/04/87 D50% = 3.25 µm 22/04/87 D50% = 1.22 µm

Fig.1 COMPARISON OF SAMPLERS (DROPLET SIZE EFFECTS)

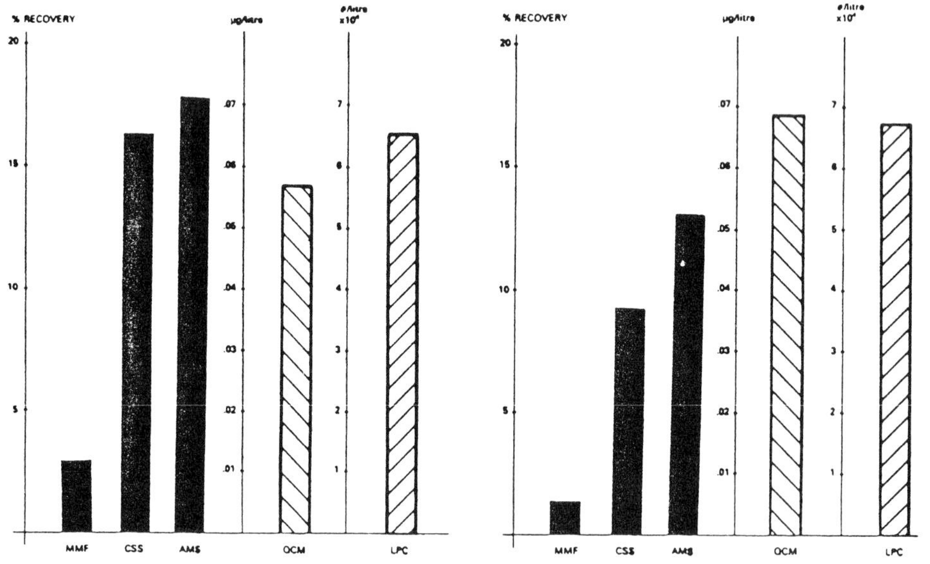

08/04/87 D50% = 3.57 µm 09/04/87 D50% = 3.49 µm

Fig.2 COMPARISON OF SAMPLERS (REPRODUCIBILITY)

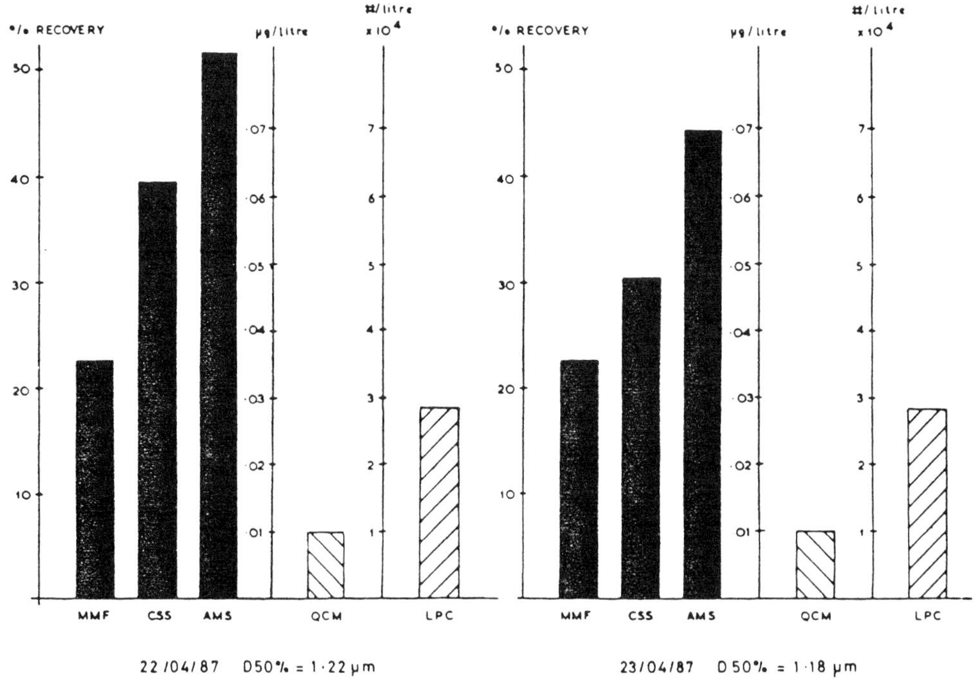

FIG 3 COMPARISON OF SAMPLERS (REPRODUCIBILITY)

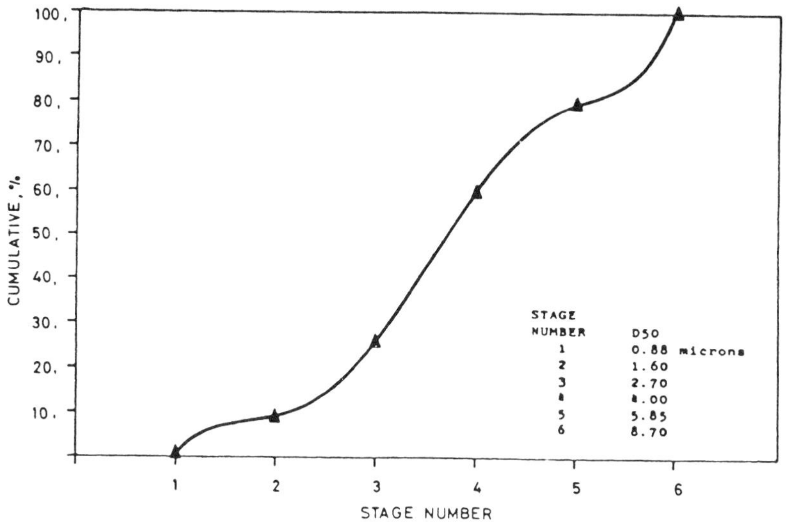

FIG 4 DROPLET SIZE DISTRIBUTION ANDERSEN MICROBIAL SAMPLER
DATE 6/4/87 D50% 3.25 MICRONS

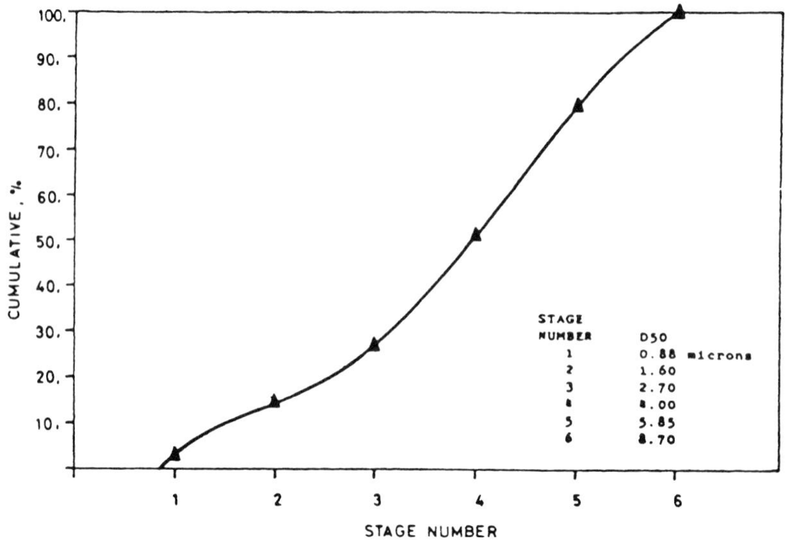

FIG. 5 DROPLET SIZE DISTRIBUTION ANDERSEN MICROBIAL SAMPLER
DATE 8/4/87 D50% 3·57 MICRONS

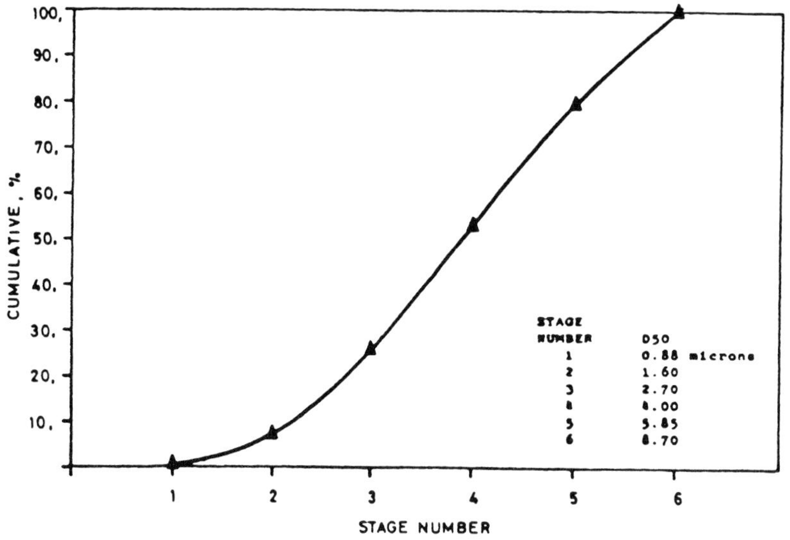

FIG. 6 DROPLET SIZE DISTRIBUTION. ANDERSEN MICROBIAL SAMPLER
DATE 9/4/87 D50% 3·49 MICRONS

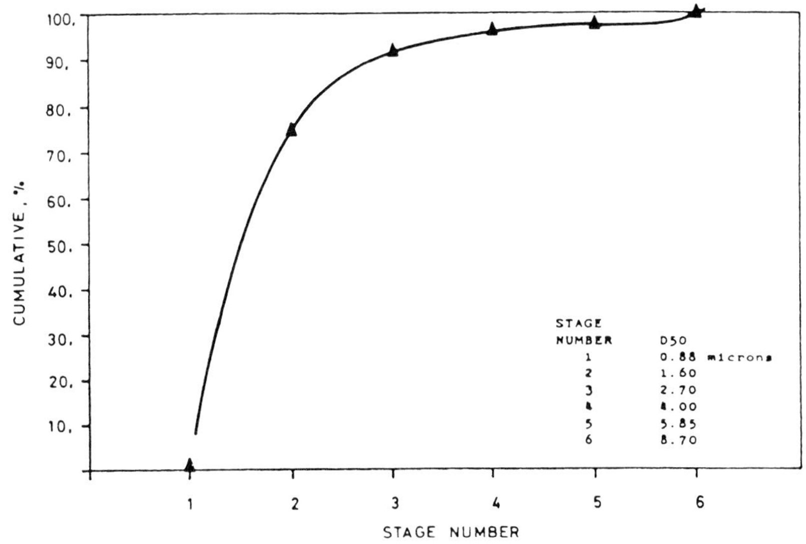

FIG.7 DROPLET SIZE DISTRIBUTION. ANDERSEN MICROBIAL SAMPLER.
DATE 22/4/87 D50% 1·22 MICRONS

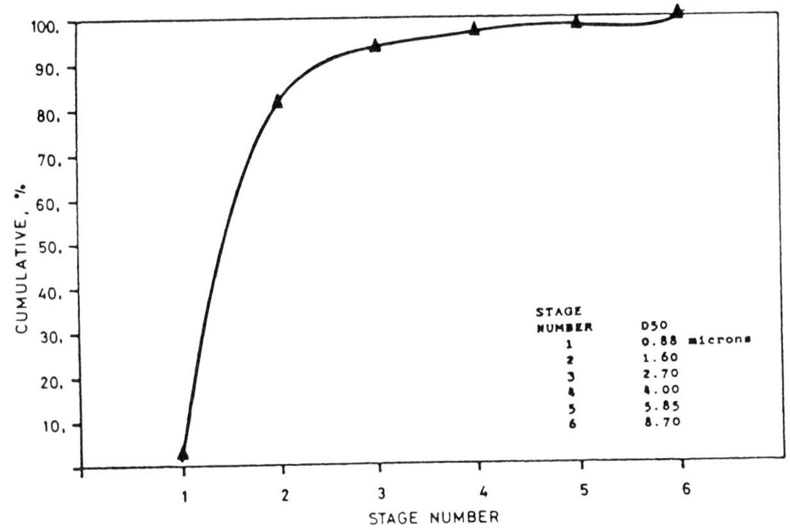

FIG.8 DROPLET SIZE DISTRIBUTION. ANDERSEN MICROBIAL SAMPLER.
DATE 23/4/87 D50% 1·18 MICRONS

Discussion

When comparing the performance of the viable aerobiological monitors the highest percentage recoveries were around 55% recorded by the AMS. Losses in percentage recoveries could be attributed to a number of factors. Firstly the droplet size, as indicated by increases in the mean droplet size decreasing the percentage recoveries for the viable samplers. Thus in a larger droplet, two or more spores may be present which impact upon the agar to form only one colony. The second possible factor is that spores from different droplets may impact at the same point and hence be combined into one colony count. [According to (4) it should be possible to distinguish colonies arising from separate spores at an early stage in the growth. To date we have not achieved this.] The phenomena of multiple impaction will be particularly significant if the collection area is relatively small such as with the MMF. Thus the trends in the percentage recovery results could be attributed to the total impaction surface. For the AMS a correction factor is given in the operating manual to account for this effect. No such correction was made for the CSS or MMF. A third possibility is the absence of colony counts in the droplets below 0.8 micron as suggested by the AMS and QCM droplet size distribution results. It has been suggested (5) that the sub-micron droplets did not contain spores despite a feed spore size of approximately 0.5 micron. The efficiency of the AMS in collecting sub-micron droplets has also been questioned. These latest possibilities will be examined in future work by sampling the exhaust air from the viable monitors.

The effect of droplet size on reproducibility also encompasses the above possibilities.

Other factors which may account for losses in % recovery include losses of droplets to the apparatus walls and loss of viability of the spores from injection to collection.

Conclusions and Recommendations

In order to have confidence in the data provided by the viable samplers used in this study, it is necessary to understand the droplet size and impaction behaviour. Such understanding will enable better interpretation of the exposure and penetration of biological emission into the human respiratory system as well as obtaining reliable and meaningful data on containment of bioprocess plant.

The problems of low percentage recovery in the impaction-type viable samplers could be overcome by first applying a correction factor for droplets containing multiple spores according to the particle size distribution, and secondly a correction that allows for the probability of multiple impaction of droplets at the same site. Alternatively impinger-type monitors using liquid media should not suffer from these problems, although other problems such as dilution errors are anticipated.

The development of these correction factors and the comparison of impactors with impingers is currently under investigation at our laboratory.

In addition, one of the activities of the new Industrial Biosafety Club (6) concerned with all aspects of bio-process plant containment, will be an R&D programme to monitor the containment of generic process components including valves and seals using suitable detection devices.

The viable sampling devices described above normally require an incubation time of 24 to 48 hours before a result is obtained. The advantage of a particle counting technique is the rapid detection of aerosols, but the disadvantage is that the composition of the aerosol is unknown. In order to develop a technique which can be used to validate process equipment, we are planning to combine both types of monitor by automatic viable sampling controlled from the particle counter.

Acknowledgement

This work forms part of a contract BAP 110 UK(H) on Risk Assessment in Biotechnology of the European Commission Biotechnology Action Programme. Financial support from the UK Health and Safety Executive is gratefully acknowledged.

References

1. OECD (1986) "Recombinant DNA Safety Considerations" Organisation for Economic Co-operation and Development Publications, Paris.

2. Gibbs, D.F. (1987) "The safer handling of active powders". Chapter 39 in Verrall, M.S. and Hudson, M.J. (Eds) "Separations for Biotechnology". Ellis Horwood Ltd, Chichester, UK.

3. Schaeffer, P. et al, (1963) Colloques Int. Cen. Matn. Rech. Scient., Marseille, 124, 553-563.

4. Anon, (1984) "Operating Manual for Anderson Sampler, Inc; Viable (Microbial) Particle Sizing Sampling". Nov, Atlanta, Georgia, USA.

5. Stewart, I.W., Watson, R.H. and Leaver, G. (1987) "Evaluation of aerobiological monitors". Chapter 37 in Verrall, M.S. and Hudson, M.J. (Eds) "Separations for Biotechnology". Ellis Horwood Ltd, Chichester, UK.

6. Norton, M., Leaver, G. and Melling, J. (1987) "Safety Aspects of Bioprocessing: The Implications of the Industrial Biosafety Club". The World Biotech Report 1987, Vol. 1, Part 1, pp 51-54, Online Publications, Pinner, UK.

STUDIES ON THE FATE AND GENETIC STABILITY OF RECOMBINANT MICROORGANISMS IN MODEL ECOSYSTEMS

Schmidt, F.R.J., Henschke, R. and Nücken, E.
Institut für Bodenbiologie, Bundesforschungsanstalt für Landwirtschaft, Bundesallee 50, 3300 Braunschweig, FRG

Summary: Risk assessment studies concerning deliberate release of recombinant microorganisms were performed in micro-ecosystems. For this purpose chemostate and microcosm model systems were designed comprising plain soil, rhizosphere and Lumbricus systems. Monitoring methods had been established to follow the fate of recombinant bugs and its genetic information. Unexpected hazardous situations following release of recombinant bugs could not be detected so far. Thus single cases of natural in situ horizontal gene transfers were characterized in detail. Their significance in ecological risk assessment studies was evaluated in these model systems. These include in vivo gene integrations involving site specific recombination directed by hot spots of transposable elements.
Keywords: Risk assessment, microcosms, rhizosphere, horizontal gene transfer, transposons, recombination.

Introduction

Recently proposals had been developed to evaluate the potential risks from deliberate release of genetically engineered organisms to the environment (1), (2), including experimental approaches (3). The sharp official debates and political furore that followed single releases of bacteria in the USA, which had been manipulated in vitro, or in Germany of drug resistant Rhizobia altered in vivo (4) raised the question, how to find the right test instrumentarium for evaluating potential ecological risks of recombinant bugs for the environment, i. e. short and long-term influence on the resident flora and fauna. Due to the set up of regulations for the release problem, a step-by-step procedure will be prescribed prior to open field testings (1). One of these tests require the use of model ecosystems as a first instrumentarium to evaluate risks. They should help to define and characterize the single parameters influencing growth, survival, stability and fate of the recombinant microorganism (5) and its DNA after release. The problem reveals the

Risk Assessment for Deliberate Releases
Edited by W. Klingmüller
© Springer-Verlag Berlin Heidelberg 1988

need of improved techniques for identifying organisms and their genetic information and to evaluate the risk of in situ gene transfer by suitable monitoring methods.

In this paper our efforts in designing micro-ecosystems and their use in risk assessment methodologies were described. Finally exemplary mechanisms directing horizontal gene transfer are introduced and considered carefully with respect to gene interactions between microbial communities in soil.

Materials and methods

Plasmids are listed in Table 1. Bacterial strains used for risk assessment studies or as plasmid hosts are: E. coli: HB101 (strA F^- recA pro leu lacZ hsdR hsdm supE (6), SK1592 (gal thi $T1^r$ sbcB15 hsdR hsdm endoA) (7), RRIΔM15 (leu pro thi strA hsdR hsdM lacZΔM15/F^+ $lacI^QZ\Delta M15$ pro^+) (8), JC2926 (recA thi thr arg his leu mal Rif^r)(9); Pseudomonas putida 2082, (R. Helmuth, Berlin), P. fluorescens DSM 50148, DSM 1694, Alcaligenes eutrophus DSM 531; Klebsiella pneumoniae DSM 30104; Enterobacter aerogenes DSM 30035; Azorhizobium caulonodans ORS 571 (Cb^r)(10); Agrobacterium tumefaciens C58 (Rif^r, pTI)(11), C58C1 (Rif^r)(11).

Soils used in microcosms were sampled from two agricultural plots at Braunschweig which remain under current characterization (12, 13). Soil texture and main clay minerals were SiCl (sand 47.1%, silt 46,2%, clay 6,7%) and kaolin minerals; bulk densities 1.2 $g \cdot cm^{-3}$. They mainly differ by their biomass content (13). These were soil I from field plots of the Agricultural Research Centre at Braunschweig, a luvisol from loess (parabrown earth): bulk density 1.2 $g \cdot cm^{-3}$, C_t 0.80%, N_t 0.08 %, pH (KCl) 5.4 and soil II from Jerxheim, a phaeozem from loess: bulk density C_t 2.45%, N_t 0.20%, pH7.4 (12). Soil manipulations were according to Anderson and Domsch (14).

Inoculation and isolation of soil microorganisms: Inoculations were according to Jagnow (15), separation of bacteria from soil was performed as previously described (16, 17, 18).

Bacterial conjugation, transformation, media and drugs were as described (19,20).

DNA techniques: Isolation, purification, electron microscopy, heteroduplex, restriction enzyme analysis, hybridization and blotting, gene cloning, transposition and other manipulations were as described earlier (20, 21).

Nucleotide sequence determination and computer analysis was carried out as described (22).

Table 1. Plasmids

Plasmids	Relevant markers	Source/Ref.
RSF1010	Sm^r Su^r	(23)
RP4	Ap^r Tc^r Km^r Nm^r Liv^r	(24)
pUC18	Ap^r	(25)
K5	Ap^r Km^r Nm^r But^r (contains S35 promoter, pA from CaMV and aphB (Tn5) in pUC18	B. Gronenborn FRG
pPCV002	Ap^r Cb^r Km^r Nm^r But^r	(26)
pHS1	Tc^r	(27)
pFL8	Sm^r Spe^r	this paper
pFL9	Sm^r Spe^r Gm^r Db^r Km^r Tb^r Sis^r TnpA TnpR	"
pFL10	Ap^r Sm^r Spe^r Cm^r Gm^r Km^r TnpA TnpR	"
pFL11	Ap^r Gm^r Db^r Km^r Tb^r Sis^r TnpA TnpR	"
pFL12	Ak^r But^r Dm^r Km^r Net^r Nm^r Tb^r Sis^r TnpA TnpR	"
pFL13	But^r Dm^r Km^r Net^r Nm^r Tb^r Sis^r TnpA TnpR	"
pFL15	Ap^r TnpA TnpR	"
pFL212	Ap^r Cm^r Tc^r TnpA TnpR	(20)
pBP201	Ap^r Sm^r Spe^r Gm^r Db^r Km^r Tb^r Sis^r TnpA (Hg^s $TnpR^-$ derivative of Tn4000)	(20)
pFL213	TnpA (Tn21)	this paper

RESULTS AND DISCUSSION

Risk assessment studies and design of micro-ecosystems

Our prime rationale for use of micro-ecosystems is simplification of natural environmental variability, which is often too large and complex that it prevents in-depth risk assessment of the structure and functioning of natural ecosystems. For this purpose we first designed microcosms of different characters. They were conducted as controlled-environmental studies under naturally fluctuating environmental conditions due to the definition of Draggan (28):"Microcosms are experimental units designed to contain important components and to exhibit important processes occurring in a whole ecosystem" and they " are functionally similar to but may differ in origin or in structure from the natural ecosystem that is simulated". One of these models had to simulate the active transport of genetically engineered microorganisms (GEM) in selected, well defined agricultural soils. It contained a column filled up with soil under defined pressure. Under these conditions Lumbricus terrestris was forced to dig its own gangway and picked up food enriched with GEM from the top and left its excrements on top of the column that were easy to isolate and characterize for its content of GEMs and their DNA. The second category of microcosms was based on well defined agricultural soils, some of which contained plants like tobacco, Brassica rappa and Sesbania rostrata, the tropical host of A. caulonodans which itself is a diazotrophic bacterium capable of inducing aerial stem and root nodules and also of fixing N_2 in the free living state and using the fixed nitrogen (NH_4^+) as primary source (10). These abilities not only facilitate the easy identification of the microorganism, but allowed to construct a microcosm with few parameters: the bug and its host grown up in sterile sand.

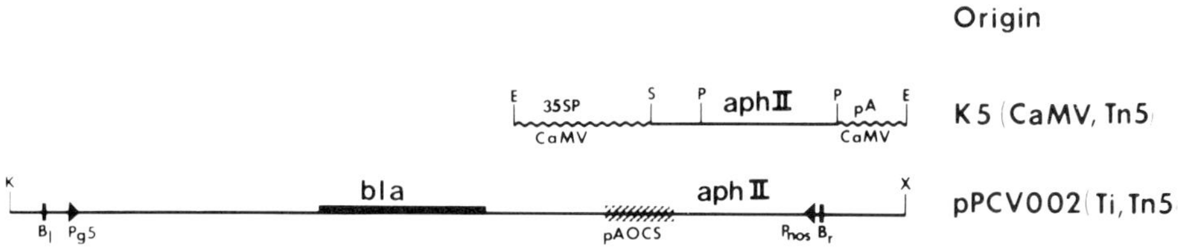

Fig. 1. Recombinant fragments with known sequence.pNOS: promoter of nopaline-, pAOCS: polyadenylation sequence of octopine synthase gene, B_l, B_r: left - and right border sequences of T-DNA.

Representative fragments (Fig. 1) were cloned into wide host range replicons RSF1010 (Tra^-) and RP4 (Tra^+) which then were transformed to several microorganisms: E. coli, Pseudomonas, Klebsiella, Enterobacter, Agrobacterium and Azorhizobium prior to release into the respective micro-ecosystems. GEMs and indigenous bacteria were identified after varying intervals by direct plating detection. While E. coli decreased below the direct plating detection limit (20-50 CFU/g soil) only slight decline in CFU was observed for the other GEMs after 10 days (10^7 - 2 x 10^4 CFU/g soil). Differences in numbers were due to content of soil, dependent on the presence of plants and the addition of nutrients. DNA transfer was monitored by hybridization with suitable restriction fragments or synthetic oligonucleotides as probes. Some of these experiments are still under evaluation. The in situ rate of transfer of RP4 derived replicons to bacteria obtained from soil ranged from 8 x 10^{-7} to 2 x 10^{-9} transferants/recipient. The higher transfer rates yielded from non-GEMs added to the microcosms whereas the lower rates were observed with indigenous soil bacteria. As monitored by selection on the respective drugs or by colony hybridization strains of E. coli, P. putida, P. fluorescens and A. tumefaciens harbouring RSF1010 hybrids (Fig. 1) were not able to transfer their genes to Enterobacter, Klebsiella, Alcaligenes, or other bacteria from plant associated microcosms under the given conditions so far. The evaluation of possible DNA transfer to the single plants is still in work and not yet ready to be presented. However our experimental conditions may not be sufficient to identify rare events of transfer. Thus we used the chemostate model to study the interaction between the single GEM (K5, pPCV002, TnpA from Tn21) and a selected collection of soil representatives as proposed by Omenn (3) under several culture conditions. Due to the actual state these experiments do not allow to present conclusions yet. However we have to improve our methods for extracting and identifying microorganisms from model systems and raise the limits for detection of putative gene transfer events. What are these limits?

Gene transfer of aminoglycoside resistance

Recently Slater (29) has summed up the present state of gene transfer in bacterial soil communities, in which conjugal transfer by plasmids play the major role. But our present knowledge about plasmids in soil microorganisms and their function is limited. To exclude conjugal transfer GEMs are required that carry their new genetic information integrated on Tra^- replicons or on the chromosome. Could this genetic information disseminate in soil habitats, too? To answer this question it is necessary to look at other mechanisms involved in gene pickup and transfer first. The dissemination of a certain class of genes could provide some information about this phenomenon. Genes conferring resistance to aminoglycosides (Ag) in pro- and eucaryotes (e.g. neomycin-phosphotransferase gene from Tn5) belong to this class. Others and our own experimental data suppose their origin in the antibiotic producers, one of which is Streptomyces kanamyceticus harbouring a gene greatly homologous to the respective one in hospitalized $Gram^-$ bacteria like E. coli, Serratia, Pseudomonas (21). Our present knowledge of the spread of these genes is summarized in Table 1. Most of the single steps in transfer had been identified including their fundamental mechanisms, their participating plasmids and class II-transposons. Single routes of dissemination by these elements had been exhibited as well as the process immobilizing transposition functions leading to stable inheritance of the resistance information in hospitalized bugs (Fig. 2).

Table 1. EVOLUTION AND DISSEMINATION OF DRUG RESISTANCE GENES

1. ORIGIN: Drug producing microorganisms (e.g. Str. kanamyceticus) Function: Self-protection from autotoxicity (Davies & Smith, 1978)

2. PRIMARY TRANSFER: Horizontal gene transfer in soil ecosystems: rhizosphere. Infection chain: Actinomycetacetae → Pseudomonadaceae → Enterobacteriaceae

3. INTEGRATION: into site specific receptor-sequences (e.g. hs1, hs2) of unstable mobile elements (Tn21)

4. SECONDARY TRANSFER:
   A) "Pre-antibiotic era": transposition into IncFII-plasmids (Datta & Hughes, 1983)
   B) "Antibiotic era": conjugal transfer of IncFII-plasmids with transposons

5. STABILE INHERITANCE: Selection of hospitalized bacteria with R-genes carried by stable immobilized genetic elements localized on: Plasmids (e.g. of Klebsiella, Pseudomonas) (Fig. 2) or Chromosomes (Serratia) (Schmidt, 1987)

There is a need of information about the original function and the mechanism of primary transfer from actinomycetes to other soil bacteria. As the Gm^r gene from Str. kanamyceticus is localized on the chromosome of its host, the vector directing its transfer in soil has to be identified. Obviously mobile genetic elements are involved in this process, but now one step in dissemination requires improved clarification, the integration of Ag^r-genes into these elements.

Gene integration adjacent to Sm^r gene of Tn21

Tn21 belongs to a class of elements with a wide host range in $Gram^-$ and $Gram^+$ bacteria. Some of these elements lack or carry additional genes (Fig. 2). Deletions and insertions were identified at the ends of the Sm^r only. The nucleotide se-

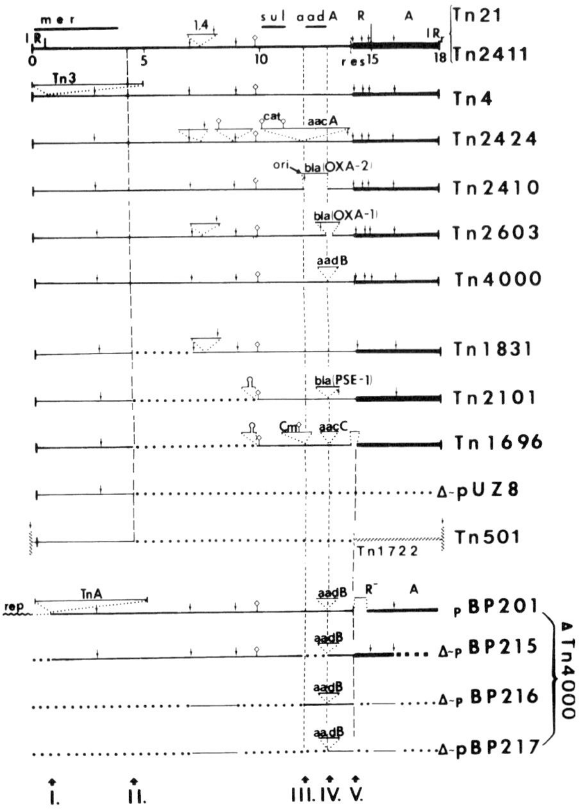

Fig. 2. Tn21 transposon family (20) with hot spots for integration, substitution and deletion: I. Tn3 transposition in mer; II. deletions adjacent to mer; III., IV. insertions, substitutions (triangles), deletions (closed circles) at hs1 and hs2 of aadA; V. substitutions at res (internal resolution site). Arrow: EcoRI, square: SmaI

quence of these hot spots for site specific recombination has been determined from some of the available elements (22). As depicted in Fig. 3 they have several common features: (i) hs2 is a deletion derivative of hs1; (ii) hs1* presents itself as an enlarged hs1 which hardly is congruent with the consensus sequence of hs1; (iii) part of the hs1 sequences include palindromic sequences with functions as rho-independent transcription terminator or Shine-Dalgarno site due to its respective position and sequence (22); (iv) hs1 is predominantly located at the N-, hs2 at the C-terminus and hs1* at the lefthand end of the insertion (22).

Gene transfer from Gram$^+$ members of soil habitats to Gram$^-$ bacteria.

How frequent are these small entities in procaryotes and what is their function in the integration process? For this purpose we have constructed a "gene integration assay" which allows to analyse integration of genes into these elements or derivatives (Fig. 4). Isolated from chromosomal loci of Gram$^-$ species (Fig. 4) these genes did not exhibit any detectable homology with Tn21 and were cloned

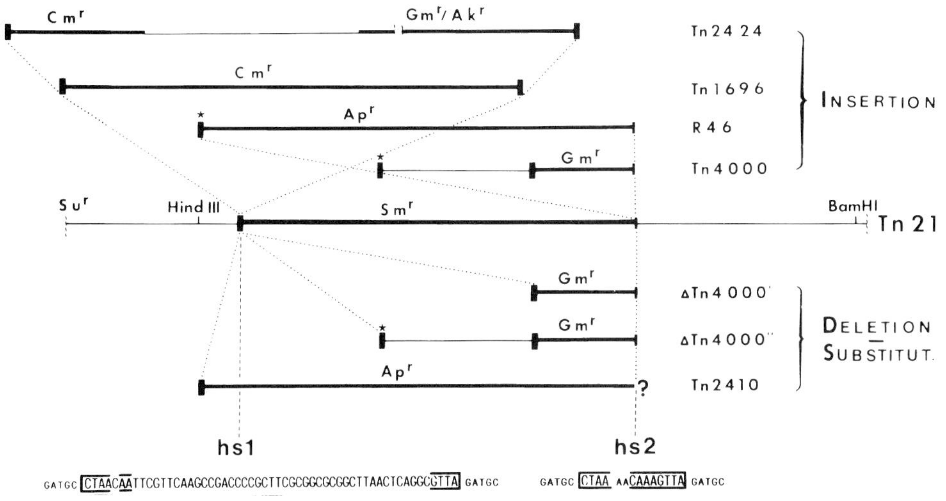

Fig. 3. Positions and consensus sequence of hot spots hs1, hs1* and hs2 in Tn21 related elements as present in the central region (Su^r, Sm^r) of each transposon.

into pHS1, a derivative of pSC101 and temperature sensitive in its replication functions (27). These experiments were performed in the chemostate with recombination pro- and deficient mutants of E. coli under varying conditions. So far integration could be observed in E. coli SK 1592 with a frequency of 10^{-8} following a temperature shift to 42°C in order to eliminate replicon pHS1. Recombinants could not yet be found in recA and recF mutants at the given level of detection (10^{-10}). Respective experiments with other rec^- derivatives and with Pseudomonas species are under evaluation. Tn21 recombinant derivatives from resistant clones were analysed by restriction, heteroduplex (Fig. 5) and sequence analysis. As deletions at hs1 or hs2 were observed with frequencies of 10^{-5}, the integration frequency under these conditions can be regarded as low. However such a frequency could be sufficient for the transfer of genetic material from a GEM after deliberate release into the open field. recA-dependent homologous recombination in E. coli requires a minimum of ≥ 20 base pairs of complete homology; however mismatches in larger segments dramatically decrease the frequency of recombination (30). Thus an illegitimate recombination mechanism has to be assumed for gene integration at hs1 and hs2 involving a new recombinase system which directs site-specific gene integration at these elements and their derivatives. In addition such integration events in P. fluorescens and other soil habitants are under evaluation to follow the gene transfer proposed in Table 1.

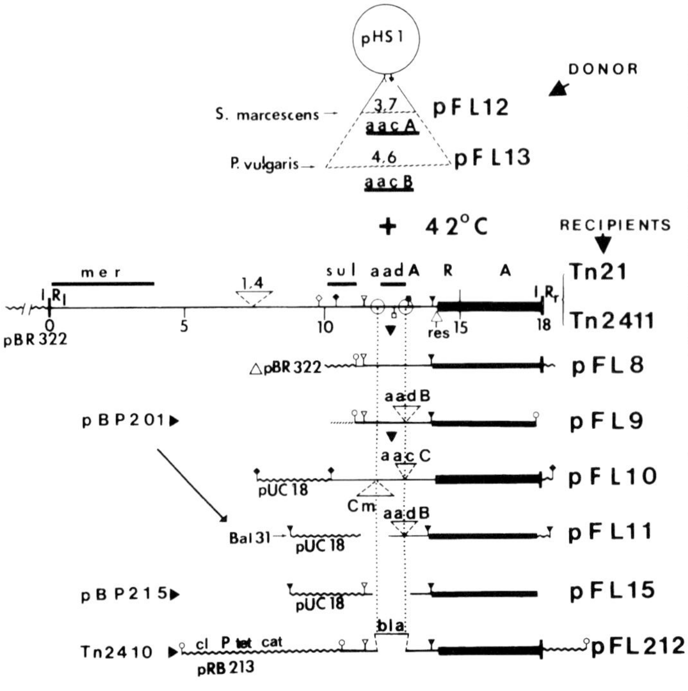

Fig. 4. Gene "integration assay". Restriction sites are symbolized.

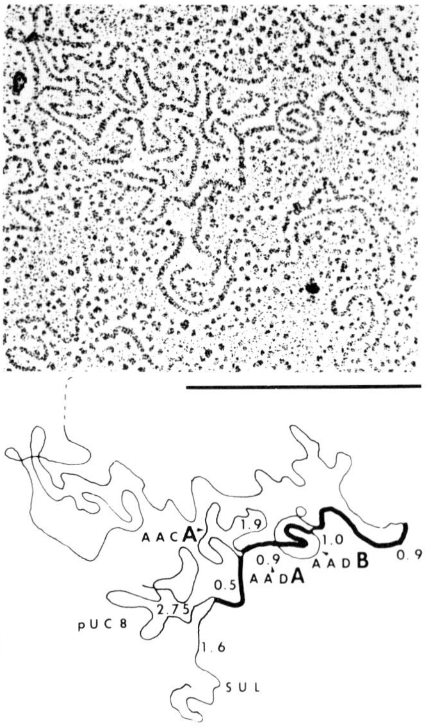

Fig. 5. Heteroduplex between Tn21 and pFL9::aacA. Bar: 0.5 µm.

In summary our approach in microbial, ecological and genetic studies revealed current need in risk assessment research concerned with the design of microcosms, growth, survival, identification, extraction and monitoring of GEMs and the associated soil flora. Furthermore characterization of plasmids, transposons and insertion elements of indigenous soil bacteria combined with detailed molecular studies in single cases of horizontal gene transfer will help to understand communication processes within soil habitats.

Acknowledgements

We thank J. Schell, B. Gronenborn and F. de Bruin for gifts of strains and A. Berger for excellent technical assistance. This work was supported by the Bundesministerium für Forschung und Technologie and the Deutsche Forschungsgemeinschaft.

REFERENCES

1. COVELLO, V.T. and FIKSEL, J.R. (1985). Finally report to the office of Science and Technology Policy; Executive Office of the President. NSF, Washington, D.C.
2. DOMSCH,K.H.; DRIESEL, A.J.; GOEBEL, W.; ANDERSCH, W.; LINDENMAIER, W.; LOTZ, W.; REBER, H. and SCHMIDT, F. (1987). Forum Mikrobiologie, in press.
3. OMENN,G.S. (1985). G1-G28. In: Covello, V.T. and Fiksel,J.R.; see 1.
4. DICKMAN, S. (1987). Nature 328, 568.
5. GILLETT, J.W. (1986). Environ. Management 10, 515-532.
6. BOYER, H.W. and ROULLAND-DUSSOIX, D. (1969). J.Mol.Biol. 41, 459-472.
7. KUSHNER, S.R. (1978). In: Proceedings of the International Symposium of Genetic Engineering. H.W.Boyer and S.Nicosia (eds.) Elsevier/North Holland. Publishing Co., Amsterdam, 17-23
8. RÜTHER, U. (1982). Nucl. Acids Res. 10, 5765-5772.
9. BACHMAN, B.J. (1972). Bacteriol. Rev. 36, 525-527.
10. PAWLOWSKI, K., RATET, P. and DE BRUIN, F. (1987). Mol. Gen. Genet. 206, 207-219.
11. HOLSTERS, M., SILVA, B., VAN VLIET, GENETELLO, C., DE BLOCK, M., DHAESE, P., DEPICKER, A., INZE, D., ENGLER, G., VILLAROEL, R., VAN MONTAGUE, M. and SCHELL, J.(1980). Plasmid 3, 212-230.
12. ANDERSON, T.-H. and DOMSCH, K.H. (1986). Z. Pflanzenernaehr. Bodenk. 149, 457-468.
13. MARUMOTO, T., ANDERSON, J.P.E. and DOMSCH, K.H. (1982). Soil Biol.Biochem. 14, 469-475.
14. ANDERSON, T.-H. and DOMSCH, K.H. (1985). Biol. Fert. Soils 1, 81-89.
15. JAGNOW, G. (1985). In: W. Klingmüller (ed). Azospirillum III: Genetics, physiology, ecology. Springer-Verlag, Berlin, Heidelberg, New York, Tokio, 203-214
16. BAKKEN,L.R. (1985). Appl. Environ. Microbiol. 49, 1482-1487
17. GODWIN,D. and SLATER,J.H. (1979). J. Gen. Microbiol. 111, 201-210
18. SAYLER, G.S., SHIELDS, M.S., TEDFORD, E.T., BREEN,A., HOOPER, S.W., SIROTKIN, K.M. and DAVIS, J.W. (1985). Appl. Environ. Microbiol. 49, 1295-1303.
19. KRATZ, J., SCHMIDT, F. and WIEDEMANN, B. (1983). J. Bacteriol. 155, 1333-1342.
20. SCHMIDT, F. and KLOPFER-KAUL, I. (1984). Mol. Gen. Genet. 197, 109-119.
21. SCHMIDT, F.R.J. (1987). J. Gen. Microbiol., submitted for publication.
22. SCHMIDT, F.R.J., NÜCKEN, E. and HENSCHKE, R. (1987). Mol. Gen. Genet., submitted for publication.
23. GUERRY, P., VAN EMBDEN, J.D.A. and FALKOW, S. (1974). J. Bacteriol. 117, 619-630.
24. JACOB, A.E., SHAPIRO, J.A., YAMAMOTO, L., SMITH, D.I., COHEN, S.N. and BERG, D. (1977). 607-638. In: Bukhari, A.I., Shapiro, J.A. and Adhya, S.L. (eds.). DNA insertion elements, plasmids, and episomes. Cold Spring Harbor Laboratory, Cold Spring Harbor, N.Y., 607-638.
25. NORRANDER,J., KEMPE, T. and MESSING, J. (1983). Gene 26, 101-106.
26. KONCZ, C. and SCHELL, J. (1986). Mol. Gen. Genet. 204, 383-396.
27. HASHIMOTO, T. and SEKIGUCHI, M. (1976). J. Bacteriol. 127, 1561-1563.
28. DRAGGAN,S. and REISA, J.J. (1980). In: J.P. Giesy, Jr. (ed). Microcosms in ecological research. Technical Information Center U.S. Dep. of Energy, iii-xii.
29. SLATER, J.H. (1985). In: Engineered organisms in the environment. Halverson, H.O., Pramer, D. and Rogul, M. (eds.). ASM, Washington, D.C., 89-98.
30. WATT, V.M., INGLES, C.J., URDEA, M.S. and RUTTER, W.J. (1985). Proc. Natl. Acad. Sci. USA 82, 4768-4772.

GENETIC VARIATION AND HORIZONTAL GENE TRANSFER: PROSPECTS OF RESEARCH OF A GROUP INSTALLED AT THE "BIOLOGISCHE BUNDESANSTALT"

Horst Backhaus and Jörg Landsmann
Institut für Biochemie, Biologische Bundesanstalt,
Messeweg 11/12, D-3300 Braunschweig

Summary: A way to define a relevant concept for research on risk assessment of deliberate releases of genetically engineered microorganisms is described. Established risk assessment schemes are used to highlight problems and information needs. The impact on decision processes is discussed.

Introduction

A group of scientists has been established at the Biologische Bundesanstalt (BBA) with the assignment of research on risk assessment of deliberate releases of genetically engineered organisms. The installation of this group is part of a research programme initiated by the Ministry of Science and Technology following recommandations of the Enquete Commission of the Deutscher Bundestag on "chances and risks of gene technology". The BBA will be a consultant institution in decisions on applications for releases of genetically engineered organisms into the environment. Such releases will be exemptions from a general ban if a proposal of a five year moratorium put forward by the commission will be followed.

As the group "Gene Technology and Safety in the Environment" has only just been created, our efforts in establishing a relevant research programme concentrate on the collection of data and questions and inspection of existing risk assessment schemes and guidelines. The lack of knowledge in the field of biological risk assessment is significant. Our task in defining our research concept is now, to screen for problems which are on the one hand most pressing to be solved and on the other hand susceptible to scientific analysis, depending on our current knowledge and techniques.

Risk Assessment for Deliberate Releases
Edited by W. Klingmüller
© Springer-Verlag Berlin Heidelberg 1988

A simple model

A brief statement of data and the elaborate models on risk assessment in radiation protection may be used to highlight the differences and problems we are confronted with when trying to apply a similar methodology to risk assessment of a deliberate release of genetically engineered organisms (1).

As mentioned by M.Alexander (this vol.), we usually differentiate the areas of - exposure and - effects assessment. Exposure assessment tries to calculate whether and to what extent endpoint organisms are exposed. In the case of radioprotection we are concerned exclusively with humans or their organs.

In order to get a quantitative evaluation we use models for

- release, which calculate probability and amount of accidental releases (this is irrelevant in case of deliberate releases because we should have precise data about the time, geographical location and number of organisms released).
- transport along distances and the - fate of the substances.

Models for transport and fate only have to take into account physical forces (wind, gravity) and physicochemical attributes, e.g. the partitioning in different media. Calculating the fate in the biosphere is more complex but data about enrichments in organisms or biologcal half lives can be obtained in experiments with low doses, in model systems, and through observations after previous accidental releases. A very sensitive detection technology is available for radioisotopes.

For the assessment of effects of an exposure to radiation, observed and experimentally obtained dose-response curves allow predictions of effects on human health. But even in this simple case of a single hazardous agent (with few physico-chemical descriptors) released , there are a variety of problems in the public and scientific debate preventing a broad consensus on risk assessment. Among these the issues of longterm effects of low doses on human health, the problems of extrapolation of results from model experiments, and the calculation and acceptance of the risks of accidental releases are worth mentioning.

Special problems of deliberate releases of organisms

Genetically engineered organisms will show a much more complex interference with their environments as do any newly introduced organisms. New and so far unsolved problems result from organism characteristics of replication in suitable environments (habitat selection and reproduction).

The reproduction of organisms but also their persistence and the survival of their products as spores (and pollen) is primarily related to exposure assessment because a predictive capability about the time, location and degree of reproduction is needed for a quantitative estimate. Our efforts to gain data and experiences about these figures depend on the detection technology: concentrations of microorganisms in a given environment might fall temporarily below a current detection level but might recover through favorable seasons or other changes in the environment(2).

The crucial point in any discussions about biological risk assessment however is the assessment of effects. Quantitative dose-effect relations can be derived if we regard toxicity or pathogenicity. In some cases the effects of human, animal or plant pathogens are well known. To mention an example: 35 bacteria of Bacillus anthracis in the blood system of humans are sufficient to cause a lethal infection. In many cases, however, we have only inferior knowledge about the mechanism of pathogenicity and host - pathogen interactions, which drastically reduces our potential to predict the effects of new combinations of genetic traits. For a comprehensive prognosis of effects on the environment the knowledge of potential host ranges would be necessary. Most of these questions have been presented time and again (e.g.3) but still, the lack of knowledge in this field is significant.

All other potential hazards could be regarded as ecological effects, which are mainly a consequence of the characteristics of habitat selection and reproduction of the organism. Only a knowledge of these parameters will make predictable effects like the potential to compete with native organisms for the same ecological niche, uncontrolled reproduction, and a hazardous imbalance of nutrient cycles.

Genetic variation and horizontal gene transfer

Additional complications for risk assessments in this field arise from the fact that organisms are subject to mutational changes and information transfer across species borders. Mutations can alter the function and expression of genes. The transfer of genes between different species of microorganisms (horizontal gene transfer) will lead to new reproduction and expression patterns in the new genetic background. These features combined with the ecological aspects will make it most difficult to gain predictive capabilities for risk assessment.

In discussions on potential risks of a release of genetically engineered organisms, mainly in relation to the problems of generating new pathogenicities or ecological disturbances, we adopt the general principle "Focus on the product (organisms) not on the process (by which it has been engineered)". Then, it should not make any difference whether a new trait has been introduced into an organism by genetic engineering or by conventional breeding.

Present transformation technology does not enable us, to place the transforming genes at a desired locus, exept for insertional mutagenesis by homologous recombination in procaryotic chromosomes or the eucaryotic yeasts. Thus, transformed genes usually have to accomodate themselves at regions in the chromosome which have not been selected for by evolution.

Mutagenic primary lesions are not evenly distributed over chromosomes, neither are fixed mutations . Also, repair processes in many cases have regional or even sequence specificities; this is also true for recombination events: cells somehow exert control over the distribution of mutational events. It then might not be a mere speculation that introduced genes could differ from other genes of their host in their susceptibility to mutational changes.

A variation in chromosomal location , which may be regarded as a mutational event in a wider sense, leads over to gene transfer. A new phenotype after horizontal transfer of genetic information does relate to both, fate and effects assessment. There are hints in the literature that the mechanism of horizontal gene transfer played a role in evolution, e.g. in parasite(symbiont) - host interactions. Within the time scale open to experimental testing or direct observation the oncogenic transformation of plants with Agrobacterium or the transformation of oncogenes with retroviruses are examples in eucaryotes. Horizontal exchange between eucaryotes, however is not usually amenable to risk assessment. This is totally different in procaryotes. For microorganisms the mechanisms of conjugation (DNA transfer by direct cell contact), transformation (uptake of naked DNA), and transduction (bacteriophages as vectors) are well known in the laboratory. Plasmids as well as chromosomal loci can be transferred and become stably integrated in the recipient`s genome. Biological specificities at different stages of the transfer mechanisms define the respective potentials to cross the borders between microbial species. Data are available that such mechanisms do occur in nature. Their relative contribution to horizontal gene transfer in natural microbial ecosystems is unknown, however. We may vizualize microbial systems (in soil, wastewater etc) as populations of different species exchanging not only products of metabolism but genes in addition. Our figure represents this view of a microbe population exchanging genetic information.

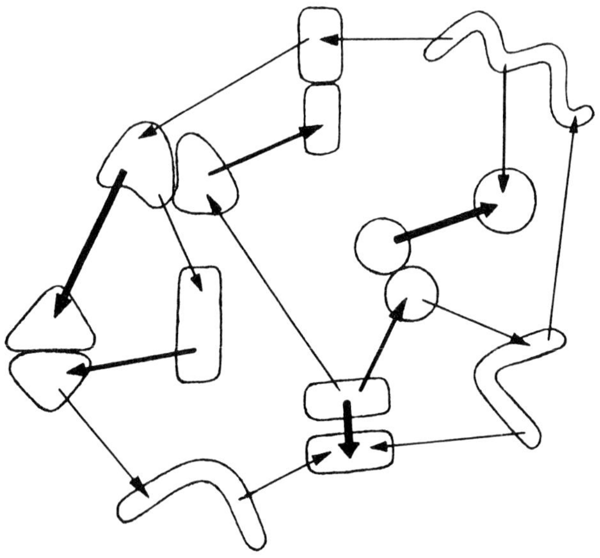

Gene flux in a microbial community

A concept for research

A concept for risk assessment research should be established between two extremes: the ultimate goal to demonstrate and quantify the probability of all potential risks or the practice of an alibi function in scientifically corroborating the insignificant risks of gene technology applications. It is rational to presume that each new technology creates its specific risks. Risk assessment research will have to provoke the worst cases in a controlled manner. This does not necessarily mean to construct AIDS viruses but, e.g. to try and show whether a gene given a beneficial and desired trait in one organism could create a hazardous phenotype in another background or after mutation or transform a nonpathogen by chance into a pathogen or a crop plant into a weedy relative.

Our own expertise includes the characterization of pathogenicity determinants of a plant pathogen (Pseudomonas syringae). This research can be expanded to questions about host range and host specificity. Host specificity studies could also be conducted on Baculo-viruses where cooperation with those groups using these viruses for plant pest control will be sought. The questions of variability and cross protection of plant viruses are already partially been tackled by other groups within the BBA.

We shall test the relationship between genotype and ecological distribution with a selected case (variants of Bacillus thuringiensis enterotoxin).

Genetic variation and horizontal gene transfer are among the main subjects of basic research applicable to risk assessment of genetically altered organisms.

In microorganisms we will try to determine the frequency and possibility of a given marker gene to become part of a moveable element. By screening for the stability of introduced genes we will try to observe differences in somaclonal variation of transformed plant cell cultures for transformed genes compared to normal genes. Techniques to tackle such questions are available: they include the search for RFLP's and the demonstration of more subtle sequence changes by other methods including genomic sequencing.

Specific processes could be studied in model ecosystems. E.g., to determine a potential contribution of the mechanism of generalized transduction to gene transfer in nature we might design a system to monitor this phage mediated gene transfer. Phage production (induction of prophages) can be stimulated by mutagens or other stresses. Avoiding cell contacts between donors and recipients will exclude conjugation and using DNAses in the experiments will allow to distinguish between transduction and DNA transformation.

If we are lucky we could measure the transfer from a specific donor to chosen recipients in a given system. Such data cannot be easily integrated in a scheme for general risk assessment. We still need more comprehensive data. But it seems impossible to observe the total gene flux in a model system after introduction of a new gene (via microorganisms, phage or DNA), and to monitor the transfer of genes to all possible recipients irrespective of gene expression, simply because of the restrictions through organism numbers, time, and detection technology. A potentially better way to obtain relevant data about the pathways and frequencies of gene flow in microbial communities might be the evaluation of "epidemiological" data on the spread of "new" genes in selective response to new environmental stresses. Antibiotic resistances and metabolic pathways for xenobiotics are examples for such newly selected genetic traits.

Horizontal gene transfer from Agrobacterium to plants may normally be a unilateral process, but to find out, whether this transfer is principally reversible we will design experiments where this might accidentally be possible (through homologous recombination or transposition).

Relevant points for deliberate releases

For the assessment of release associated risks quantitative risk estimates are needed. Where experimental data are unobtainable risk probabilities have to be added up point by point and educated guesses described explicitly. To date it is impossible to deline a generally applicable scheme for risk assessment. Assessments must be made case by case. Catalogues of questions have to be put forward and updated with the increase in knowledge and experience. Answers to some critical questions will become compulsory in any applications for deliberate releases.

What is known about the survival and reproduction of the GEM (genetically engineered microorganism)?

Habitat selection potential is not predictable from physiological or other data obtained in the laboratory, including the genetic manipulation itself. The argument of an additional "metabolic burden" due to an added genetic trait will not generally be applicable.

What is known about the ecological distribution of the host organism?

The release of microorganisms about which we posess a fair amount of ecological data (e.g. for Rhizobium from longterm applications in agriculture) will be prefered over "exotic" organisms.

Where is the new gene located?

We cannot generally exclude the possibility of horizontal gene transfer but we might influence the mobility of genes through special constructions (by integration of the gene into the chromosome instead of using plasmids which might be mobilizeable or have a broad host range).

What is known about the phenotype of the introduced gene?

Because we have so little predictive capabilities about the potential of distribution any indications of a possible harmful interference with the ecosystem (toxicity, pathogenicity etc) has to be considered seriously.

Gene technology and the public

Along with the exposure of the environment and the human population with the products of gene technology comes the need for information. A debate about potential risks left alone to scientists, will not create automatically a public understanding and acceptance. Misunderstanding and disapprobations might turn the anticipated economical gains into large losses.

It is up to the scientist to present the new gene technology with its positive and negative aspects in a way the layman can comprehend. We can demonstrate openly and clearly those cases where genetic variation and gene transfer have resulted in new varieties, new host ranges and new chromosome numbers in nature.

But we also can describe those instances where a supposedly well characterized gene has led to surprise discoveries about their action later on. After fears and facts will be put into perspective the public will determine the level of risk it agrees to accept as a consequence of the benefits of applied gene technology.

References

(1) Vincent T.Covello, J.R.Fiksel, Eds. (1985): The Suitability and Applicability of Risk Assessment Methods for Environmental Applications of Biotechnology. Final report to the Office of Science and Technology Policy. National Science Foundation, Washington, D.C.

(2) Douglas McCormick (1986).Bio/Technology 4, 419-422

(3) Harlyn O.Halvorson, D.Pramer, M.Rogul, Eds. (1985): Engineered Organisms in the Environment: Scientific Issues. American Society for Microbiology, Washington,D.C.

UK EXPERIENCE IN REGULATING THE RELEASE OF GENETICALLY MANIPULATED MICROORGANISMS

Beringer, J.E.
Depart. of Microbiology, Univ. of Bristol, The Medical School,
University Walk, Bristol BS8 1 TD, U.K.

In the United Kingdom laboratory based genetic manipulation experiments are regulated by the Health and Safety Executive (HSE) and its Advisory Committee on Genetic Manipulation (ACGM). There is legal enforcement of safety standards for genetic manipulation by HSE through the Health and Safety at Work Act, which imposes general duties on employers to maintain a safe place of work and not to put non-employees at risk. This Act also requires that within the work place or laboratory individual workers have a duty towards the safety of others. An important role of the ACGM is to provide guidelines to enable genetic manipulation local safety committees (which are mandatory) and the scientists proposing to do research to determine appropriate containment levels for the work. These guidelines are also an expression of how to satisfy the general duties of the Health and Safety at Work Act. ACGM's primary role is to advise HSE (and the Health and Safety Commission) with respect to the general duties of the Health and Safety at Work Act which is concerned with human health and safety. However the Committee can also advise government ministers such as those for the environment, agriculture and health.

It was thought necessary to set up a Planned Released Sub-Committee of ACGM to draw up guidelines and regulate experiments involving the release of genetically manipulated organisms into the

Risk Assessment for Deliberate Releases
Edited by W. Klingmüller
© Springer-Verlag Berlin Heidelberg 1988

environment. The Planned Release Sub-Committee reports to the main ACGM Committee but operates independently of it when assessing release proposals. Some members of the Planned Release Sub-Committee are also members of the main Committee. Because there has been very little experience world-wide of the release of genetically manipulated organisms into the environment, the sub-committee was established on the basis that it would be necessary to assess proposals on a case-by-case basis and that guidelines would have to be very general in nature to allow flexibility and to ensure that important, but presently unrecognised, aspects of risk would be considered. It was not, and is still not, thought possible to prepare a detailed check list of questions. The ability of the sub-committee to operate on a case-by-case basis without a detailed check list is constrained by the number of applications coming to it, because it is time consuming. At present there are very few proposals and there is, as yet, no suggestion that numbers will increase rapidly over the next few years. It is hoped that experience gained in the UK and other countries can be used to draw up guidelines in the future which will enable certain categories of experiment, which have already been subject to exhaustive examination, to be handled much more quickly.

The membership of the Planned Release Sub-Committee (Table 1) reflects a need to balance academic expertise with an input from existing bodies with interests in human health, agriculture and the environment. In a number of cases genetically manipulated organisms will be used as biological control agents or human food and therefore will be subject to assessment by other bodies involved in product based approval schemes for pesticides, food etc. Great care is taken to ensure that relevant bodies see proposals to determine whether they also have an interest. Scientists wishing to release genetically manipulated organisms and members of the sub-committee use the ACGM/HSE/Note 3 "The planned release of genetically manipulated organisms for agricultural and environmental purposes; guidelines for risk assessment and for the notification of proposals for such work", as a base for preparing a case for release and for assessing risks. Important points from these guidelines are summarised in Table 2. It is clear from this summary that some questions are not relevant to all releases and for many types of release complete answers are not, and may never be, available.

A major problem for the United Kingdom and all other regulatory bodies is the need for an unambiguous definition of genetic manipulation. In the U.K. the Planned Release Sub-Committee decided that existing definitions used for laboratory-contained experiments were too narrow given our present state of ignorance, and that it would be useful to start with a broad definition which could be narrowed later when more experience was available. This definition (Table 3) includes organisms produced by conjugation, which is the reason why the E.E.C.-funded experiment on the release of Rhizobium (see Chapters) was assessed in the U.K. even though it was not thought to be a genetically manipulated organism in France and Germany. This definition and all other aspects of the guidelines are subject to constant revision in light of experience.

To date the Sub-Committee has assessed and approved five proposals (Table 4) all of which were for experiments that involved the release of limited numbers of organisms into trial plots. As yet there have been no proposals for large scale releases or for approval to market manipulated organisms. The procedure used is outlined in Table 5. It will be seen from this that much emphasis is placed on continuing contact between the proposers and the Planned Release Sub-Committee. This system provides a very flexible method for assessing proposals and for developing expertise. However, it is not a system that would work efficiently if large numbers of applications were forthcoming. Experience gained in the U.K. and other countries over the next few years should lead to a co-ordination of procedures and a streamlining of assessment committees.

Table 1
Constitution of the ACGM Planned Release Sub-committee

Academics (ca. 8) chosen to reflect relevant ecological and genetical expertise
Forestry Commission (1)
Environmental Health Officer (1)
Trade Unions (1)
Industry (1)
Nature Conservancy Council (2)
Department of Trade and Industry (1)
Ministry of Agriculture, Fisheries and Food (3)
Department of Health and Social Security (1)
Health and Safety Executive (1)
Department of the Environment (1)
National Environmental Research Council (1)
Secretariat (HSE, 2)

Most representatives are scientists with relevant expertise. The committee is chaired by an academic.

Table 2

Abbreviated guidelines

(a) Notification

The notifier should be advised by a local body which should include relevant scientific expertise.

Notification should be send to the Health and Safety Executive which will liaise with other relevant bodies (e.g. Pesticide Safety).

The notifier should address the points listed in the guidelines, provide any other relevant information and include details of the aims and rationale behind the project.

The HSE secretariat would like to initiate a dialogue with professors at an early stage to facilitate the preparation of proposals.

(b) Risk assessment factors.

General points to take into account:

1. The nature of the organism (e.g. species, pathogenicity).
2. The procedure used to introduce the genetic modification.
3. The nature of any altered nucleic acid and its source, its intended function/purpose and the extent to which it has been characterised.
4. Verification of the genetic structure of the novel organism.
5. Genetic stability of the novel organism.
6. Effects of the manipulation may be predicted to have on the behaviour of the organism in its natural habitat.

7. The ability of the organism to form long-term survival forms, eg spores, seeds, etc., and the effect the altered nucleic acid may have on this ability.

8. Details of any target biota (eg pest in the case of a pest control agent); known effects of non-manipulated organism and predicted effect of manipulated organism.

Information required about the method for release:

---

1. Geographical location, size and nature of the site of release and, physical biological proximity to man and other significant biota. In the case of plants, proximity to plants which may be cross pollinated.

3. Method and amount of release, rate frequency and duration of application.

4. Monitoring capabilities and intentions; how may novel organisms be traced, eg to measure effectiveness of application.

5. On-site worker safety procedures and facilities.

6. Contingency plans in event of unanticipated effects of novel organism.

Information required about the ability of the released organism to survive and spread in the environment.

---

1. Growth and survival characteristics of the host organism and the effect the manipulation may have.

2. Susceptibility to temperature, humidity, dessiccation, UV etc, and ecological stress.

3. Details of any modification to the organism designed to affect its ability to survive and to transfer genetic material.

4. Potential for transfer of inserted DNA to other organisms including methods for monitoring survival and transfer.
5. Methods to control or eliminate any superfluous organism or nucleic acid surviving in the environment or possibly in a product.

Requests for copies of the guidelines and notification of release should be addressed to the Health and Safety Executive (MD A2), Baynards House, 1 Chepstow Place, London W2 4TF.

Table 3

The definition of genetic manipulation

Genetic manipulation means the formation of new combinations of heritable material by insertion of nucleic acid molecules, produced by whatever means outside the cell into any virus, bacterial plasmid, or other vector system so as to allow their incorporation into a host organism in which they do not naturally occur, but in which they are capable of continued propagation.

Health and Safety (Genetic Manipulation) Regulations 1978.

The Planned Release guidelines include the definition above together with the following:

Organisms constructed by techniques that involve the exchange of genetic information between species (such as cell fusion, conjugation, micro-injection, and micro-encapsulation) which may otherwise fall outside the scope of the above definition. The use of potentially infective nucleic acid molecules and organisms that have undergone intentional gene deletion is also included.

The guidelines do not apply to organisms produced by classical methods of strain improvement.

Table 4

Release examined by the Planned Release Sub-Committee.

1. National Environmental Research Council (NERC), Institute of Virology, Oxford. The release of a baculovirus containing a short sequence of non-coding DNA, to facilitate epidemiological studies

Restricted to small, netted, field plot.

2. Agricultural and Food Research Council , Rothamsted Experimental Station, Harpenden.

(a) The release of a <u>Rhizobium</u> strain carrying the drug-resistance transposon Tn<u>5</u>, to act as a marker for ecological studies and gene transfer to other bacteria.

Restricted to small plot.

(b) The growth of potato plants derived from a cell fusion experiment between two species of potato, to examine progeny for lines which may have desirable properties for potato breeding programmes (e.g. disease resistance).

Restricted to trial plot and controls to prevent pollen on and formatithe survival of tubers.

3. AFRC, Plant Breeding Instutute, Cambridge.

The growth of potato plants carrying bacterial genes (beta-glucuronidase and neomycin phosphotransterase) torced to plant promoters, to study the expression of foreign genes under "natural" conditions.

Restricted to trial plot and controls to prevent pollen formation and the survival of tubers.

5. NERC, Institute of Virology, Oxford.

The release of a baculovirus modified so that the viruses are no longer protected within a protein body, to determine whether such viruses might be desirable inoculants because of a decreased ability to survive in the environment.

Restricted to a small, netted field plot.

Table 5

Assessment procedure.

1. Preparation of proposal and consideration by local genetic manipulation safety committee. Informal contact with ACGM secretariat encouraged.
2. Submission of proposal to ACGM secretariat and circulation to members.
3. Meeting of Planned Release Sub-Committee. Proposor(s) invited to give oral evidence and answer questions from the committee.
4. Written response to proposor(s), usually involving requests for further information.
5. One, or if necessary, more meetings of the Planned Release Sub-Committee.
6. Written agreement to the project sent to proposors with a request to provide the Sub-Committee with report on the work. Announcement to the press and other relevant organisations who can help keep the public informed.

LEGAL ASPECTS OF RISKS IN RELEASING GENETICALLY ENGINEERED MICRO-ORGANISMS, WITH SPECIAL EMPHASIS ON THE PROTECTION OF INDUSTRIAL PROPERTY

Dr. Erich Häusser

Deutsches Patentamt, Zweibrückenstraße 12, D-8000 München 2

1. Issues

No achievement of technology so far has been a greater challenge to our ethical and moral values than that posed by genetic engineering. The risk of its techniques not only destroying the biological balance on our planet, but also the social and ethic foundations of our society, cannot be belittled. On the other hand, the results of this research area are associated with justified hopes for a satisfactory solution to a number of medical, social and biological problems of our time.

The rapid development of genetic engineering generates new tasks and problems for legislation, but also for government agencies and the courts. An answer must be found to the question of whether and how to protect new developments in the field of genetic engineering, for instance by patents or comparable industrial property rights. On the other side, concern arises as to how humans, their property, and the environment may be effectively protected against the potential hazards of genetic engineering, and how comprehensive compensation may be secured for damages caused by a release.

Risk Assessment for Deliberate Releases
Edited by W. Klingmüller
© Springer-Verlag Berlin Heidelberg 1988

2. Protection against Risks Involved in Genetic Engineering

a) Constitutional Aspects

The different opinions supported in the scientific world as to the risks involved in the release of genetically modified micro-organisms agree at least in that detrimental impacts of a release cannot be totally excluded from the start. Under these circumstances, the legislator, who is under the obligation to protect third parties' life, health and property as well as the environment, will have to address the question of how this obligation can be fulfilled in an adequate form.

A general prohibition of all or certain release experiments seems to be to far-reaching. Taking into account the rapid technical advance in the field of genetic engineering, scientists may well be expected soon to be able to reliably predict the course of individual release experiments. Prohibition of a release experiment that has proved to be harmless, would not be indicated by the constitution. Besides, it would be an obstacle to new scientific findings and, thus, to advance in technology. In the presence of a not generally excluded threat to objects of constitutional protection, on the one side, and with the anticipated possibility of proving the safety of a release, on the other, introduction of mandatory approval for release experiments as recommended by the Commission of Inquiry of the German Bundestag, appears a legally justified and balanced solution. It has been demonstrated in the United States, the presently leading nation in the field of genetic engineering, that introduction of an obligation to obtain approval does not inevitably lead to an obstruction to technological advance. But it might be a suitable instrument to reduce general reservations and concern vis-à-vis genetic engineering, and to improve its ambient field.

b) Regulatory Aspects

So far, the regulatory laws of the Länder alone are directed against disturbance of public safety and peace by the release of genetically modified micro-organisms. In addition thereto, any individual may, in the case of release, institute the negatory action based on Sec. 1004 of the Civil Code. The two instruments, however, are not satisfactory because, as a rule, the nonspecialized regulatory agencies as well as individuals are lacking the expertise required to assess the hazardous or harmless nature of a release. In contrast, introduction of a reservation as to admissibility combines the advantage of an effective control of the hazards with the benefit for research workers and scientists of being able to carry out release experiments after an investigation by competent public authorities on a safe legal basis, a permission.

In addition, introduction of binding safety provisions, including, if necessary, penal provisions, will have to be considered as a further preventive measure. These could be oriented according to the "Richtlinien zum Schutz vor Gefahren durch in vitro neu kombinierte Nukleinsäuren" (Guidelines for the protection against risks from in vitro recombined nucleic acids) as submitted by the Federal Minister of Research and Technology (BMFT) in the presently 5th edition of May 28, 1986; compliance with the Guidelines, however, is mandatory only for all search and development projects directly or indirectly subsidized by the government, while voluntary compliance is expected for the rest.

c) Liability Aspects

In the Civil Code, the German legislator has made the principle of fault the normal case of liability. Legal liability may presently be presumed to exist on the part of those who release genetically engineered micro-organisms only in the event of negligence in handling the micro-organism which causes harm to the health or property of third parties. In view of the standards to be observed with regard to the duty to care, the safety requirements, as provided for in the BMFT Guidelines, could be applied.

The liability in tort has been tightened up by the courts on a case-by-case basis as the so-called legal duties to maintain safety have been developed. A duty to maintain safety is incumbent upon the person who creates or keeps up a risk, and consequently also upon those who work with genetically modified micro-organisms. Although liability in the case of infringement of the duty to maintain safety presupposes presence of a fault, too, the latter is easier to evidence because the person creating the source of danger is presumed to be aware of the risks and the means of control, and to use them.

In the case of observance of the recognized rules of science and technology, those who release the micro-organisms have no liability for damages under the provisions governing the liability in tort. The legislator will have to examine whether the shifting of the remaining risks to third parties, which factually results therefrom, is appropriate or whether statutory definition of strict liability is called for. The solution supported by the Commission of Inquiry according to which strict liability should be introduced in the areas of genetic engineering where the remaining risks are still high due to the absence of a secured state of science and technology, while maintaining the liability in tort in other fields, seems adequate and balanced.

Strict liability should be associated with compulsory coverage to guarantee discharge of the liability for damages. The obligatory insurance which could be modelled on the relevant provisions of the Atomic Energy Act of July 15, 1985, would not present financing problems for smaller enterprises if, following the example of the nuclear and the pharmaceutical industries, the possibility of insuring in an insurance pool was provided for.

d) Industrial Property Aspects

As far as patent applications are concerned, the goal is to effectively counter inadvertent release of micro-organisms by the parties to the procedure. In this respect, in particular availability to any third party of a deposited micro-organism from the laying-open of the patent application, as provided for by the Patent Law, involves a safety risk. This risk could be reduced on the one hand by introducing mandatory safety regulations for the handling of micro-organisms, and on the other hand, by special provisions within the existing rules on the deposit which could be oriented according to the rules on the processing of applications which have a state secret as subject matter.

A further possibility to be considered is to release deposited micro-organisms only to neutral experts sufficiently acquainted with the relevant safety provisions. This measure is likely to create an ideal balance between the public's needs both for information and security without entailing the disadvantages for the inventor which would result from an introduction of secrecy provisions.

3. Patent Protection for Inventions in the Field of Genetic Engineering

Both the German Patent Law and the European Patent Convention admit microbiological processes and micro-organisms per se to patent protection. Sec. 2(2) 2nd sentence of the Patent Law whose text and systematics are almost identical with Art. 53

of the European Patent Convention, expressly excludes microbiological processes or the products thereof from the exceptions to patentabtility as applicable to plant or animal varieties or essentially biological processes for the production of plants or animals under Sec. 2(2) first sentence of the Patent Law.

Purpose and function of the Patent Law serve as an indication that government oversight of the risks and hazards genetic engineering involves is not to be set up in this legal area. The patent, as a technical protective right limited in time, derives its legitimation, under legal policy and legal philosophy aspects, from the constitutional guarantee of intellectual property (Art. 14 of the Basic Law), but also from the aim of the State to promote technical advance. The grant of a patent which gives the inventor the exclusive right to industrial exploitation of his invention, constitutes both a reward for a technical solution achieved and an incentive to search for such solutions; for patent protection secures a competitive advantage for the enterprise or the inventor of the new patent, which permits recovery of the search and development costs incurred prior to the invention. In return for this exclusive right, the inventor must disclose his invention to the public. Protection of property, promotion of advance in technology, and economic incentives, these are the three main goals of patent protection.

The legislator consistently refrained from making the grant of protective rights contingent on considerations as to social usefulness or other ethic or moral values. Only if publication or exploitation of an invention were contrary to public order, i.e. the basic principles of our legal system, or against morality, would the German Patent Office as a granting authority be entitled to reject a patent application (Sec. 2(1) of the Patent Law). Inventions relating to genetic engineering and their exploitation, however, do neither always include per se a considerable hazard to the citizens' legal positions protected by the Basic Law, nor can they be rated as obviously detrimental from a societal point of view, either.

The patent grant procedure also serves the transparency and a limited ethic control of this field of research: in return for the exclusive right granted by the patent which issues at a later date, the patent owner must disclose his invention to the Patent Office, and, 18 months from the date of filing or the priority date, also to the public in a manner sufficiently clear for a person skilled in the art to carry it out. By means of the disclosure of the application and the invention therein contained, as required by the Patent Law, professional circles as well as regulatory agencies, for instance in the field of control of pharmaceutics or the area of sanitary police, are given the opportunity to inform themselves and to examine its nature and effects in detail.

Concern expressed as to the possibility that incompletely searched processes and products of genetic engineering might be put on the market and cause serious damage to the population seems unjustified. On the contrary, patentability has as a result that the inventor will strive for completeness of the invention he wishes to have patented, failing which he would risk not to be granted the protective right in the absence of patentability requirements which include the reproducibility of the invention by persons skilled in the art and, thus, completion up to the stage where the alleged effects are produced.

For the reasons mentioned, the question as to whether new developments in the field of genetic engineering may, and should be granted patent protection, should be clearly affirmed. It cannot be ignored that the consistency of a patent application with public order and morality as a requirement for the grant of the exclusive right, definitely reduces the interest in a commercial exploitation of not completely mature inventions relating to genetic engineering. Scientists and enterprises will make use of the advantage of industrial exclusive rights, developing processes and products of genetic engineering within a framework of moral and regulatory limitations to be determined yet, which will serve the health of the individual, the food supply of all, the protection of nature and, thus, the general welfare.

THE APPROACH OF THE U.S. ENVIRONMENTAL PROTECTION AGENCY IN REGULATING CERTAIN BIOTECHNOLOGY PRODUCTS

Elizabeth A. Milewski Ph. D.

U.S. Environmental Protection Agency
401 M. Street, S.W.
Washington D.C. 20460

The U.S. Environmental Protection Agency functions under a number of statutes to carry out its mission of protecting human health and the environment. While these statutes were not written specifically for biotechnology or products of biotechnology, there are two which have been interpreted as investing EPA with the authority to do so. These are the Federal Insecticide, Fungicide and Rodenticide Act (FIFRA) and the Toxic Substances Control Act (TSCA).

FIFRA creates a statutory framework under which EPA regulates, by a registration statute, the sale and distribution of pesticides. A pesticide will be registered for use only if EPA determines that the pesticide will not cause or increase the risk of unreasonable adverse effects to humans or the environment. In order to register a pesticide, an applicant must submit or cite to EPA data on subjects such as product composition, toxicity, environmental fate, and effects on nontarget organisms. This process involves premarket review of data on a pesticide's safety, and regulation of the use of a pesticide before it is registered, since some of the data needed for a registration review must be developed in field tests under actual use conditions. In order to test in the field, an applicant may obtain an experimental use permit authorizing limited use of an unregistered product. Such a permit may be issued if the proposed experiment will generate data needed to register a pesticide and the experiment will not cause unreasonable adverse effects on the environment.

FIFRA gives EPA oversight responsibility over microbial pest control agents whether or not these agents are genetically engineered. The fact that EPA has the authority to approve pesticides for testing or use as a prerequisite for entry into the environment is a powerful regulatory tool. FIFRA places on the registrant or applicant the burden of proof that the benefits of use of the product outweigh the risks.

TSCA authorizes EPA to acquire information on "chemical substances" and "mixtures" of chemical substances in order to identify potential hazards and exposures. EPA can regulate the production, processing, distribution, use and disposal of chemical substances if they present an unreasonable risk of injury to health or the environment.

Under TSCA, EPA can also require testing of any "chemical substance" that may present an unreasonable risk of injury to health or the environment or is produced in substantial quantities and may result in substantial environmental release or substantial human exposure.

Risk Assessment for Deliberate Releases
Edited by W. Klingmüller
© Springer-Verlag Berlin Heidelberg 1988

TSCA gives EPA jurisdiction over the manufacture, processing, distribution, use, and disposal of all chemicals in commerce or intended for entry into commerce that are not specifically covered by other regulatory authorities (foods, drugs, cosmetics, pesticides, etc.). Thus, TSCA serves as "gap filling" environmental law. It is concerned with all exposure media (air, water, soil, sediment, biota).

The heart of TSCA is section 5, which implements the act's goal of identifying potentially hazardous new chemical substances before they enter commerce. This section requires that manufacturers and importers of "new" chemical substances intended for commercial use submit a premanufacture notice to EPA at least 90 days before manufacture or import is to begin. EPA has 90 days to prohibit or regulate the production, processing, distribution or import and disposal of a "new" substance, otherwise manufacture or import may begin. The Agency review period can be extended to 180 days if necessary for good cause. The manufacturer or importer must submit information as specified in the notification form, and which is either "known or reasonable ascertainable" at the time of submission. They must also submit test data that are in their possession and under their control at the time of submission.

EPA is using various statutory mechanisms under FIFRA and TSCA to ensure that it receives reports in advance of the first test release of any particular microorganism which falls under EPA purview.

TSCA's applicability to regulating microbial biotechnology products is based on the interpretation that microbes are chemical substances under TSCA [OSTP]. Under TSCA, microorganisms which through deliberate human intervention were altered to contain genetic material from dissimilar source organisms are considered "new" and subject to premanufacture notification (PMN) requirements under section 5 if they are manufactured or imported for TSCA purposes. EPA defines "dissimilar" as organisms from different taxonomic genera; the organisms resulting from such a combination are called "intergeneric".

In addition to the section 5 PMN requirements, two other provisions of TSCA provide key elements for oversight of microorganisms produced by the biotechnology industry. These are the section 5 significant new use authority and the section 8 reporting authority.

Recognizing that some microorganisms or applications may present less concern than others, the EPA has constructed a system with two different levels of review; more information will be evaluated in one level of review than in the other and the review period will differ.

EPA at present intends to focus regulatory emphasis on three categories of microorganisms. The three types of microorganisms that will receive the greatest regulatory emphasis are microorganisms with "new" characteristics or that are new to the environment in which they will be used (and whose behavior is therefore less predictable), microorganisms that have pathogenic attributes (pathogens or organisms containing genetic material from pathogens), and microorganisms that are used in the environment (and therefore have the potential for widespread exposure).

EPA will implement this policy under TSCA in several ways: (1) In order to require that a PMN review be conducted prior to the environmental release of any "new" microorganism, the Agency will amend an existing rule so that environmental releases in the course of "research and development" on a microorganism will not be exempt from PMN review as they are under the current PMN rule. (2) Under section 5(a)(2) of TSCA, the Agency will issue a significant new use rule (SNUR) which will cover certain microorganisms that are pathogens or have been deliberately altered to contain genetic material from pathogens. (3) All other microorganisms which are to be deliberately released to the environment will be subject to reporting requirements under TSCA section 8(a). Section 8(a) reporting requirements will specify fewer information requirements than a section 5 PMN or SNUR review will specify. EPA is requesting that industry voluntarily comply with the first two of these policies until they are actually implemented by the Agency.

As part of the experimental use permit process, EPA has developed special procedures for genetically engineered and nonindigenous microorganisms.

EPA's current policy for small scale field testing of microorganisms subject to FIFRA also provides for two levels of notification and review. The three categories of microorganisms that EPA has determined will receive the greatest regulatory emphasis will be subject to 90-day notification and review to determine if an experimental use permit is required. There is also an expedited, abbreviated, 30-day review for small scale testing of all other genetically engineered or nonindigenous microbial pesticides. At the end of the 30-day review, EPA will determine whether or not an EUP is required, or whether the experiment may proceed without further review.

Those parts of the policy under FIFRA requiring rulemaking would formally codify, as part of the Experimental Use Permit (EUP) Regulations, the Interim Policy Statement issued in October 1984 [EPA] which eliminated the exemption for field tests involving less than 10 acres of land and one acre of water for genetically engineered pesticides.

For the immediate future, EPA will review submissions received under FIFRA and TSCA on a case-by-case basis. EPA will incorporate both review by EPA scientists and review by recognized experts into the process in order to obtain the best scientific advice available. EPA emphasis on external scientific peer review is noteworthy. Recognized scientists possessing expertise pertinent in the review of genetically engineered organisms (including topic areas such as ecology, microbial ecology, molecular genetics, physiology, etc.) have been consulted on an _ad hoc_ basis. Included in this group are members of the EPA Scientific Advisory Panel, a statutory advisory group established to provide external review and comment on scientific issues related to the regulation of pesticides.

As part of EPA's efforts to obtain the best scientific advice available, EPA has instituted a Biotechnology Science Advisory Committee (BSAC). This committee, which is advisory to the Administrator of the EPA, is composed of 11 voting members: nine scientists and two representatives of the public. The BSAC will analyze problems, conduct reviews, offer recommendations to the Agency, help provide state-of-the-art scientific advice on the issues of biotechnology, and offer an additional opportunity to the public to participate in the regulatory process. Scientist members of the BSAC have been selected on the basis of their professional qualifications to examine questions of hazard, exposure, and risk to humans, other nontarget organisms, and ecosystems or ecosystem components. The BSAC will be supplemented by consultants when they are needed to extend the range of expertise and experience of the standing committee.

In evaluating a field trial, the Agency assesses hazard and exposure and integrates those assessments to provide a risk assessment. The hazard assessment identifies potential adverse effects, if any, of the microorganisms, and estimates the probability that these effects will occur in exposed populations and ecosystems. The exposure assessment describes potential exposures that might result from the release of the microorganisms.

Although these assessments are performed by separate teams, the hazard and exposure assessments are frequently integrated throughout the evaluation process. This integration of the two assessments throughout the process facilitates the risk assessment. Because risk is the product of the exposure and hazard probabilities, their effects can counterbalance. For example, if either of the probabilities approach zero, risk is very low.

Throughout the assessment, the Agency takes into consideration the effects of uncertainty and data sufficiency/insufficiency.

Both FIFRA and TSCA require that benefit be estimated and considered in judging the field trial. Some risk will be tolerated if there is sufficient benefit to balance the risk.

EPA has indicated [OSTP] the types of data to be submitted for review prior to small scale field testing of microbial pesticides. Similar information as appropriate for the microorganism and the proposed test would be submitted under TSCA.

The information which should be submitted for a level 1 review is:

(1) The identity of the microorganism, including characteristics, and means and limits of detection.

(2) Description of the natural habitat of the microorganism or its parental strains, including information on natural predators, parasites, and competitors.

(3) Information on the host range of the parental strain(s) or nonindigenous microorganisms.

(4) Information on the relative environmental competitiveness of the microorganism, if available.

(5) If the microorganism is genetically engineered, information should be provided on the methods used to genetically engineer the microorganism(s): the identity and location of the rearranged or inserted/deleted gene segment(s) in question; a description of the new trait(s) or characteristic(s) that are expressed; information on potential for genetic transfer and exchange with other organisms, and on genetic stability of any inserted sequences.

(6) A description of the proposed testing program, including site location, crop to be treated, target pest, amount of test material to be applied and method of application.

The information which should be submitted for a level 2 review is:

(1) Background information on the microorganism

° Identity of the microorganism, including tables of characteristics, and means and limit of detection using the most sensitive and specific methods available

° Description of the natural habitat of the microorganism or its parental strain, including information on natural predators, parasites, and competitors

° Information on host range, especially infectivity and pathogenicity to nontarget organisms

° Information on survival and ability of the microorganism to increase in numbers (biomass) in the environment (e.g., laboratory or containment facility test data)

° If the microorganism is genetically altered, the following information should be provided in addition to the information listed above

- Information on the methods used to genetically alter the microorganism

- Identity and location of the rearranged or inserted/deleted gene segment(s) in question (host source, nature, base sequence data, or restriction enzyme map of the gene(s))

- Information on the control region of the gene(s), and a description of the new trait(s) or characteristic(s) that are expressed

- Information on potential for genetic transfer and exchange with other organisms, and on genetic stability of any inserted sequence

- Information on relative environmental competitiveness compared to the parental strains.

(2) Description of the proposed field test

° The purpose or objective of the proposed testing

° A detailed description of the proposed testing program, including test parameters

° A designation of the pest organism(s) involved (common and scientific names)

° A statement of composition for the formulation to be tested, giving the name and percentage by weight of each ingredient, active and inert, production methods, contamination with extraneous microorganisms, potency and amount of any toxins present, and where applicable the number of viable microorganisms per unit weight or volume of the product (or other appropriate system for designating the quantity of active ingredient)

° The amount of pesticide product proposed for use and the method of application.

° The State(s) in which the proposed program will be conducted, and specific identification of the exact location of the test site(s) (including proximity to residences and human activities, surface water, etc.)

° The crops, fauna, flora, geographical description of sites, modes dosage rates, frequency, and situation of application on or in which the pesticide is to be used

° A comparison of the natural habitat of the microorganism with the proposed test site

° The number of acres, number of structural sites, or number of animals/plants, by State, to be treated or included in the area of experimental use, and the procedures to be used to protect the test area from intrusion by unauthorized individuals

° The proposed dates or period(s) during which the testing program is to be conducted, and the manner in which supervision of the program will be accomplished

° A description of procedures for monitoring the microorganisms within and adjacent to the test site during the field test

° The method of disposal or sanitation of plants, animals, soil, etc., that were exposed during or after the field test

° Means of evaluating potential adverse effects and methods of controlling the microorganism if detected beyond the test area.

References

[EPA] U. S, Environmental Protection Agency. 1984. Microbial Pesticides: Interim Policy on Small Scale Testing; Notice of Interim Policy. Federal Register 49:202.

[OSTP] Office of Science and Technology Policy, U.S. Environmental Protection Agency. 1986. Coordinated Framework for Regulation of Biotechnology; Announcement of Policy and Notice for Public Comment. Federal Register 51:23313-23336.

Subject Index

WIDENER UNIVERSITY
WOLFGRAM
LIBRARY
CHESTER,